C0-ALU-984

HIGH
PERFORMANCE
INSTRUMENTATION
AND AUTOMATION

HIGH PERFORMANCE INSTRUMENTATION AND AUTOMATION

Patrick H. Garrett

Taylor & Francis
Taylor & Francis Group

Boca Raton   London   New York   Singapore

A CRC title, part of the Taylor & Francis imprint, a member of the
Taylor & Francis Group, the academic division of T&F Informa plc.

Published in 2005 by
CRC Press
Taylor & Francis Group
6000 Broken Sound Parkway NW, Suite 300
Boca Raton, FL 33487-2742

© 2005 by Taylor & Francis Group, LLC
CRC Press is an imprint of Taylor & Francis Group

No claim to original U.S. Government works
Printed in the United States of America on acid-free paper
10 9 8 7 6 5 4 3 2 1

International Standard Book Number-10: 0-8493-3776-3 (Hardcover)
International Standard Book Number-13: 978-0-8493-3776-5 (Hardcover)
Library of Congress Card Number 2004030435

This book contains information obtained from authentic and highly regarded sources. Reprinted material is quoted with permission, and sources are indicated. A wide variety of references are listed. Reasonable efforts have been made to publish reliable data and information, but the author and the publisher cannot assume responsibility for the validity of all materials or for the consequences of their use.

No part of this book may be reprinted, reproduced, transmitted, or utilized in any form by any electronic, mechanical, or other means, now known or hereafter invented, including photocopying, microfilming, and recording, or in any information storage or retrieval system, without written permission from the publishers.

For permission to photocopy or use material electronically from this work, please access www.copyright.com (http://www.copyright.com/) or contact the Copyright Clearance Center, Inc. (CCC) 222 Rosewood Drive, Danvers, MA 01923, 978-750-8400. CCC is a not-for-profit organization that provides licenses and registration for a variety of users. For organizations that have been granted a photocopy license by the CCC, a separate system of payment has been arranged.

Trademark Notice: Product or corporate names may be trademarks or registered trademarks, and are used only for identification and explanation without intent to infringe.

Library of Congress Cataloging-in-Publication Data

Garrett, Patrick H.
   High performance instrumentation and automation / Patrick H. Garrett.
      p. cm.
   Includes bibliographical references.
   ISBN 0-8493-3776-3
   1. Engineering instruments. I. Title.

TA165.G36 2005
670.42'75--dc22   2004030435

Taylor & Francis Group
is the Academic Division of T&F Informa plc.

Visit the Taylor & Francis Web site at
http://www.taylorandfrancis.com

and the CRC Press Web site at
http://www.crcpress.com

Preface

This book integrates sensor-based instrumentation with process automation systems. This convergence provides increased performance by means of more homogeneous designs enabling new process applications including molecular manufacturing. That advancement is aided by innovations that have occurred in many computer-centered industry and laboratory implementations, such as data fusion attribution in multisensor information systems (described in the process automation chapters). Throughout the text, a unified modeled-error measurement and control system representation further quantifies sensed information and process parameter performance, which is also compatible with enterprise quality representations. These developments are supported by 45 tables of engineering design data.

Chapter 1 presents a compendium of process sensors of ascending complexity for mechanical, to quantum, to analytical measurements. The next three chapters address linear signal acquisition circuits and systems, including five categories of instrumentation–amplifier sensor interfaces, active filter bandlimiting, and signal conditioning quality upgrading for coherent and random interference. In Chapter 5, data conversion devices (including seven A/D converter types) and their performance are analyzed. Chapter 6 presents the theory of sampled data, employing intersample error to evaluate system design influences ranging from noise aliasing to the relationship among signal bandwidth, amplitude, and sampling rate, plus the effectiveness of various output signal interpolation functions such as closed-loop bandwidth in digital control systems.

The instrumentation system examples described in Chapter 7 include video digitization and reconstruction, integrated I/O design, mass airflow measurement, robotic axes volumetric error, and multisensor process error propagation accompanied by comprehensive model-based analysis. This development also beneficially supports an ascending process control signal hierarchy from environmental parameters to controlled variables that enhance process apparatus capabilities to analytical assays.

Automation Chapters 8 through 13 illustrate a concurrent engineering taxonomy constituting process-to-control structure design methods proofed and demonstrated by a diversity of industry and laboratory case studies. Chapter 8 introduces six generic concurrent engineering linkages to achieve manufactured product properties approximated by translated automation system variables; robustness to process disorder enhanced through apparatus design that decouples sensitive subprocess parameters; production planner modeling for ideal process state trajectory execution; system design for sufficient control-loop bandwidth to meet subprocess response requirements; physical apparatus tolerance design for achieving process parameter variability minimization; and advanced automation system methods for product quality maturity.

The remaining five chapters illustrate contemporary process automation systems, including semiconductor processing, aerospace composites manufacturing, superconductor production, ceramic densification, and steel annealing variously assisted by rule-based, fuzzy logic, and neural network-based computationally intelligent process modeling. A three-level hierarchical system architecture permits more focused sensor feedback control for environmental–*in situ–ex situ* subprocesses, including feedforward production planner direction of process variables along their ideal trajectories, to achieve required product properties and attenuate unmodeled processing disorder.

The author accepts responsibility for the developments described in this book, many of which have not appeared in other books or papers, and hopes the ideas presented stimulate further contributions to those topics. Separate problems and solutions manuals for this book are also available from the publisher.

Patrick H. Garrett

Contents

chapter one

Thermal, mechanical, quantum, and analytical sensors

1.0 Introduction

Contemporary laboratory automation, manufacturing process control, analytical instrumentation, and aerospace systems all would have diminished capabilities without the availability of computer-integrated multisensor information structures. This text develops supporting error models that enable a unified performance evaluation in the design and analysis of linear and digital instrumentation systems, including compatibility with enterprise quality requirements. These methods then serve as a quantitative framework for the design of high-performance automation systems in the following chapters.

This chapter specifically describes the front-end electrical sensor devices for a broad range of applications from industrial processes to scientific measurements. Examples include environmental sensors for temperature, pressure, level, and flow; optical sensors for measurements beyond apparatus boundaries, including spectrometers for chemical analytes; and material and biomedical assays sensed by microwave microscopy. Advancements in higher attribution sensors has meant that they are increasingly being substituted for process models in many applications.

1.1 Instrumentation error interpretation

Measured and modeled electronic device, circuit, and system error parameters are defined in this text for combination into a quantitative instrumentation performance representation for computer-centered measurement and control systems. Thus the integration and optimization of these systems may be achieved by design realizations that provide total error minimization. Total error is graphically described in Figure 1.1, and analytically expressed

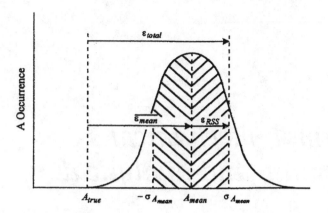

Figure 1.1 Instrumentation error interpretation.

by equation (1.1), as a composite of mean error contributions (barred) plus the root sum square (RSS) of systematic and random uncertainties:

$$\varepsilon_{total} = \sum \overline{\varepsilon_{mean}} \, \%FS + \left[\sum \varepsilon^2_{systematic} + \sum \varepsilon^2_{random} \right]^{1/2} \%FS1\sigma \qquad (1.1)$$

Total error thus constitutes the deviation of a measurement from its absolute true value, which is traceable to a standard value harbored by the National Institute of Standards and Technology (NIST). This error is traditionally expressed as 0–100% of full scale (%FS), where the RSS component represents a one-standard-deviation confidence interval and accuracy is defined as the complement of error (100% − ε_{total}). Demonstrated throughout this book is an end-to-end design goal whereby included system or device elements do not exceed 0.1%FS error contribution.

The following definitions provide an overview of relevant terms:

Accuracy: The closeness with which a measurement approaches
               the true value of a measurand, usually expressed as a
               percentage of full scale.

Error: The deviation of a measurement from the true value
               of a measurand, usually expressed as a percentage
               of full scale.

Tolerance: Allowable deviation about a reference of interest.

Precision: An expression of a measurement over some span,
               described by the number of significant figures
               available.

Resolution: An expression of the smallest quantity to which a
               quantity can be represented.

Span: An expression of the extent of a measurement between
                   any two limits.
Range: An expression of the total extent of measurement values.
Linearity: Variation in the error of a measurement with respect
                   to a specified span of the measurand.
Repeatability: Variation in the performance of the same measurement.
Stability: Variation in a measurement value with respect to a
                   specified time interval.

Sensor development has created significant technology advancements. Sensor nonlinearity, a common source of error, can be minimized by multipoint calibration. Practical implementation often requires the synthesis of a linearized sensor to achieve the best asymptotic approximation to the true value of a measurand over a measurement range of interest.

The cubic function of equation (1.2) is an effective linearizing equation demonstrated over the full 700°C range of a commonly applied Type-J thermocouple, which is tabulated in Table 1.1.

$$y = AX + BX^3 + \text{intercept} \tag{1.2}$$

Coefficient for 10.779 mV at 200°C:

$$200°C = A\ (10.779\ \text{mV}) + B\ (10.779\ \text{mV}^3) + 0°C$$

$$A = 18.5546\ \frac{°C}{mV} - B\ (116.186\ \text{mV}^2)$$

Coefficient for 27.393 mV at 500°C:

$$500°C = 508.2662°C - B\ (3182.68\ \text{mV}^3) + B\ (20{,}555.0\ \text{mV}^3)$$

$$A = 18.6099 \qquad B = -0.000475\ \frac{°C}{mV^3}$$

Table 1.1 Sensor Cubic Linearization

Y^a °C	X^b mV	y^c °C	$\varepsilon_{\%FS} = \lvert (Y - y)100\%/700°C \rvert$
0	0	0	0
100	5.269	98	0.27
200	10.779	200	0
300	16.327	302	0.25
400	21.848	401	0.23
500	27.393	500	0
600	33.102	599	0.17
700	39.132	700	0

$^a Y$ = true temperature.
$^b X$ = Type-J thermocouple signal.
$^c y$ = linearized temperature.
Note: 0.11%FS mean error; 0°C intercept; 700°C full scale.

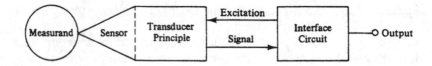

Figure 1.2 Generic sensor elements.

Solution of the *A* and *B* coefficients at judiciously spaced temperature values defines the linearizing equation with a 0°C intercept. Evaluation at linearized 100°C intervals throughout the thermocouple range reveals temperature values nominally within 1°C of their true temperatures, which corresponds to typical errors of 0.25%FS. It is also useful to express the average of discrete errors over the sensor range, producing a mean error value of 0.11%FS for the Type-J thermocouple. This example illustrates a design goal proffered throughout this text of not exceeding one-tenth percent error for any contributing system component. Extended polynomials permit further reduction in linearized sensor error while incurring increased computational burden, where a fifth-order equation can beneficially provide linearization to 0.1°C, corresponding to 0.01%FS mean error.

1.2 Temperature sensors

Thermocouples are widely used temperature sensors because of their ruggedness and broad temperature range. Two dissimilar metals are used in the Seebeck-effect temperature-to-EMF (electromotive force) junction, with transfer relationships described by Figure 1.3. Proper operation requires

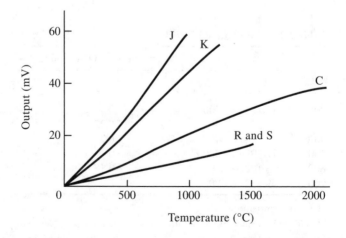

Figure 1.3 Temperature–millivolt graph for thermocouples.

Table 1.2 Thermocouple Comparison Data

Type	Elements, +/–	mV/°C	Range (°C)	Application
E	Chromel/constantan	0.063	0 to 800	High output
J	Iron/constantan	0.054	0 to 700	Reducing atmospheres
K	Chromel/alumel	0.040	0 to 1,200	Oxidizing atmospheres
R&S	Pt-Rh/platinum	0.010	0 to 1,400	Corrosive atmospheres
T	Copper/constantan	0.040	–250 to 350	Moist atmospheres
C	Tungsten/rhenium	0.012	0 to 2,000	High temperature

the use of a thermocouple reference junction in series with the measurement junction to polarize the direction of current flow and maximize the EMF measurement. Omission of the reference junction introduces an uncertainty evident as a lack of measurement repeatability equal to the ambient temperature.

An electronic reference junction that does not require an isolated supply can be realized with an Analog Devices AD590 temperature sensor, as shown in Figure 4.5. This reference junction usually is attached to an input terminal barrier strip in order to track the thermocouple-to-copper circuit connection thermally. The error signal is referenced to the Seebeck coefficients in mV/°C of Table 1.2 and provided as a compensation signal for ambient temperature variation. The single calibration trim at ambient temperature provides temperature tracking within a few tenths of a °C.

Resistance-thermometer devices (RTDs) provide greater resolution and repeatability than thermocouples, the latter typically being limited to approximately 1°C. RTDs operate on the principle of resistance change as a function of temperature, and are represented by a number of devices. The platinum resistance thermometer is frequently utilized in industrial applications because it offers accuracy with mechanical and electrical stability. Thermistors are fabricated from a sintered mixture of metal alloys forming ceramic that exhibits a significant negative temperature coefficient. Metal film resistors have an extended and more linear range than thermistors, but thermistors exhibit approximately ten times the sensitivity. RTDs require excitation, usually provided as a constant-current source, to convert their resistance change with temperature into a voltage change. Figure 1.4 presents the temperature-resistance characteristic of common RTD sensors.

Optical pyrometers are utilized for temperature measurement when sensor physical contact with a process is not feasible, but a view is available. Measurements are limited to energy emissions within the spectral response capability of the specific sensor used. A radiometric match of emissions between a calibrated reference source and the source of interest provides a current analog corresponding to temperature. Automatic pyrometers employ a servo loop to achieve this balance, as shown in Figure 1.5. Operation to 5000°C is available.

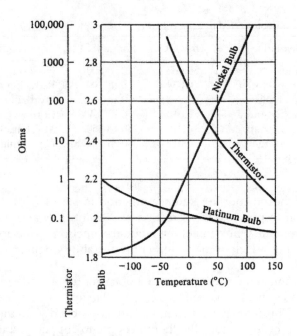

Figure 1.4 Resistance-thermometer device (RTD) temperature sensors.

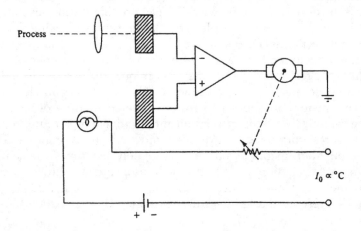

Figure 1.5 Automatic pyrometer.

1.3 Mechanical sensors

Fluid pressure is defined as the force per unit area exerted by a gas or liquid on the boundaries of a containment vessel. Pressure is a measure of the energy content of hydraulic and pneumatic (liquid and gas) fluids. Hydrostatic pressure refers to the internal pressure at any point within a liquid directly proportional to the liquid height above that point, independent of vessel shape. The static pressure of a gas refers to its potential for doing work, which does not vary uniformly with height as a consequence of its compressibility. Equation (1.3) expresses the basic relationship between pressure, volume, and temperature as the general gas law.

$$\frac{\text{Absolute pressure} \times \text{Gas volume}}{\text{Absolute temperature}} = \text{Constant} \tag{1.3}$$

Pressure typically is expressed in terms of pounds per square inch (psi) or inches of water (in H_2O) or mercury (in Hg). Absolute pressure measurements are referenced to a vacuum, whereas gauge pressure measurements are referenced to the atmosphere.

A pressure sensor responds to force and provides a proportional analog signal via a pressure–force summing device. This usually is implemented with a mechanical diaphragm and linkage to an electrical element such as a potentiometer, strain gauge, or piezoresistor. Quantities of interest associated with pressure–force summing sensors include their mass, spring constant, and natural frequency. Potentiometric elements are low in cost and have high output, but their sensitivity to vibration and mechanical nonlinearities combine to limit their utility. Unbonded strain gauges offer improvements in accuracy and stability, with errors to 0.5%FS, but their low output signal requires a preamplifier. Present developments in pressure transducers involve integral techniques to compensate for the various error sources, including crystal diaphragms for freedom from measurement hysteresis. Figure 1.6 illustrates a microsensor circuit pressure transducer for enhanced reliability, with an internal vacuum reference, chip heating to minimize temperature errors, and a piezoresistor bridge transducer circuit with on-chip signal conditioning.

Liquid levels are frequently required process measurements in tanks, pipes, and other vessels. Some of the sensing methods employed include float devices, differential pressure, ultrasonics, and bubblers. Float devices offer simplicity and various means of translating motion into a level reading. A differential-pressure transducer can also measure the height of a liquid when its specific weight, W, is known, and a ΔP cell is connected between the vessel surface and bottom. Height is provided by the ratio of $\Delta P/W$.

Accurate sensing of position, shaft angle, and linear displacement is possible with the linear variable-displacement transformer (LVDT). With this device, an AC excitation introduced through a variable-reluctance circuit is induced in an output circuit through a movable core that determines the amount of displacement. LVDT advantages include overload capability and

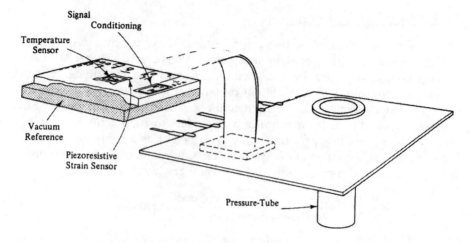

Figure 1.6 Integrated pressure microsensor.

temperature insensitivity. Sensitivity increases with excitation frequency, but a minimum ratio of 10:1 between excitation and signal frequencies is considered a practical limit. LVDT variants include the induction potentiometer, synchros, resolvers, and the microsyn. Figure 1.7 describes a basic LVDT circuit with both AC and DC outputs.

Fluid flow is measured by differential-pressure or mechanical-contact sensing. Flow rate, F, is the time rate of fluid motion, with dimensions typically in feet per second. Volumetric flow, Q, is the fluid volume per unit time, such as gallons per minute. Mass flow rate, M, for a gas is defined, for example, in terms of pounds per second. Differential-pressure flow sensing

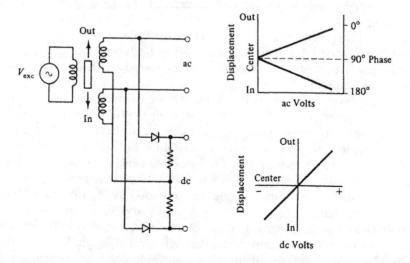

Figure 1.7 Basic linear variable-displacement transformer (LVDT) device.

elements also are known as variable-head meters because the pressure difference between the two measurements, ΔP, is equal to the head. This is
equivalent to the height of the column of a differential manometer. Flow rate
is therefore obtained with the 32 ft/sec^2 gravitational constant, g, and differential pressure by equation (1.4). Liquid flow in open channels is obtained
by head-producing devices such as flumes and weirs. Volumetric flow is
obtained with the flow cross-sectional area and the height of the flow over
a weir, as shown by Figure 1.8 and equation (1.5).

$$\text{Flow rate, } F = \sqrt{2g\Delta P} \text{ ft/sec} \qquad (1.4)$$

$$\text{Volumetric flow, } Q = \sqrt{2gL^2H^3} \text{ ft}^3/\text{sec} \qquad (1.5)$$

$$\text{Mass flow, } M = \sqrt{R\frac{\Delta P_o}{\Delta P_x}} \bullet \sqrt{\frac{P\Delta P}{T}} \text{ lb/sec} \qquad (1.6)$$

where
   R = universal gas constant
   ΔP_o = true differential pressure, $P_o - P_\infty$
   ΔP_x = calibration differential pressure

   Acceleration measurements are principally of interest for shock and
vibration sensing. Potentiometric dashpots and capacitive transducers have
largely been supplanted by piezoelectric crystals. Their equivalent circuit is
a voltage source in series with a capacitance, as shown in Figure 1.9, which
produces an output in coulombs of charge as a function of acceleration
excitation. Vibratory acceleration results in an alternating output, typically
of very small value. Several crystals are therefore stacked to increase the
transducer output. As a consequence of the small quantities of charge transferred, this transducer usually is interfaced to a low-input-bias-current
charge amplifier, which also converts the acceleration input to a velocity
signal. An ac-coupled integrator will then provide a displacement signal that
may be calibrated, for example, in millinches of displacement per volt. A
suitable integrator is shown in Figure 4.11.
   A load cell is a transducer with an output proportional to an applied force.
Strain gauge transducers provide a change in resistance due to mechanical
strain produced by a force member. Strain gauges may be based on a thin
metal wire, foil, thin films, or semiconductor elements. Adhesive-bonded
gauges are the most widely used, with a typical resistive strain element of
350 Ω that will register full-scale changes to 15 Ω. With a Wheatstone bridge
circuit, a 2-V excitation may therefore provide up to a 50-mV output signal
change, as described in Figure 1.10. Semiconductor strain gauges offer high
sensitivity at low strain levels with outputs of 200 to 400 mV. Miniature tactile
force sensors can also be fabricated from scaled-down versions of classic
transducers employing MEMS (micro-electro-mechanical system) technology.

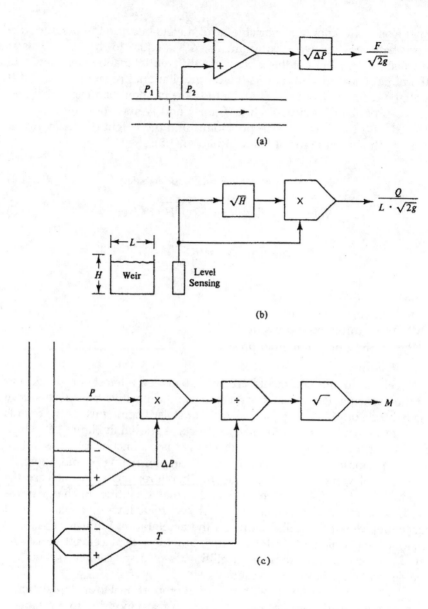

Figure 1.8 (a) Flow rate; (b) volumetric flow; (c) mass flow.

A multiplexed array of these sensors can provide sense feedback for robotic part manipulation and teleoperator actuators.

Ultrasound ranging and imaging systems are increasingly being applied for industrial and medical purposes. A basic ultrasonic system is illustrated by Figure 1.11. It consists of a phased array transducer and associated signal processing, including aperture focusing by means of time delays, and is employed for both medical ultrasound and industrial nondestructive testing.

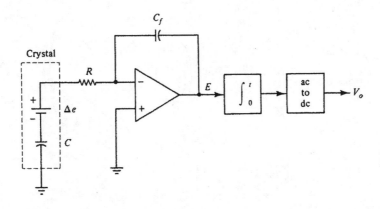

Figure 1.9 Vibration measurement transducer.

Multiple frequency emissions in the 1 to 10 MHz range are typically employed to prevent spatial multipath cancellations. B-scan ultrasonic imaging displays acoustic reflectivity for a focal plane, and C-scan imaging provides integrated volumetric reflectivity of a region around the focal plane.

Hall-effect transducers, which usually are silicon substrate devices, frequently include an integrated amplifier to provide high-level output. These devices typically offer an operating range from −40 to +150°C and a linear output. Applications include magnetic field sensing and position sensing with circuit isolation such as the Micro Switch LOHET device, which offers a 3.75-mV/Gauss response. Figure 1.12 describes the principle of Hall-effect operation. When a magnetic field, B_z, is applied perpendicular to

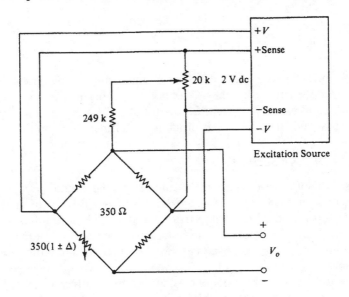

Figure 1.10 Strain gauge circuit.

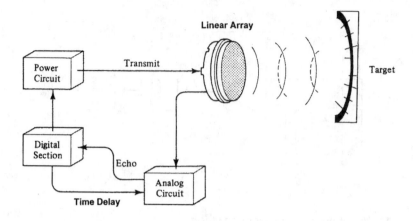

Figure 1.11 Phased array ultrasound system.

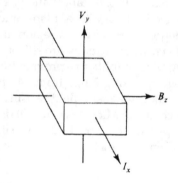

Figure 1.12 Hall-effect transducer.

a current-conducting element, a force acts on the current, I_x, creating a diversion of its flow proportional to a difference of potential. This measurable voltage, V_y, is pronounced in materials such as InSb and InAs and occurs to a useful degree in Si. The magnetic field usually is provided as a function of a measurand.

1.4 *Quantum sensors*

Quantum sensors are of significant interest as electromagnetic spectrum transducers over a frequency range extending from the far-infrared region of 10^{11} Hz through the visible spectrum about 10^{14} Hz to the far-ultraviolet region at 10^{17} Hz. These photon sensors are capable of measurements of a single photon whose energy, E, equals hv, or watt seconds in radiometry units (from Table 1.3), where h is Planck's constant of 6.62×10^{-34} Joule seconds and v frequency in Hz. Frequencies lower than infrared represent the microwave region, and those higher than ultraviolet constitute x-rays, which require different transducers for measurement. In photometry, one

Table 1.3 Quantum Sensor Units

Parameter	Radiometry	Photometry	Photonic
Energy	Watt • sec	Lumen • sec	Photon
Irradiance	Watts/cm^2	Footcandles	Photons/sec/cm^2
Flux	Watts	Lumens	Photons/sec
Area radiance	$\dfrac{\text{Watts/steradian}}{\text{cm}^2}$	Footlamberts	Photons/sec/cm^2
Point intensity	Watts/steradian	Candelas • steradian	Photons/sec/steradian

lumen represents the power flux emitted over one steradian from a source of one candela intensity. For all of these sensors, incident photons result in an electrical signal by an intermediate transduction process.

Table 1.4 describes the relative performance among principal sensors, whereby in photo diodes photons generate electron-hole pairs within the junction depletion region. For this transduction process, photo transistors offer signal gain at the source over the basic photo diode. In photoconductive cells, photons generate carriers that lower the sensor bulk resistance, but a restricted frequency response limits their utility. These sensors are shown by Figures 1.13 and 1.14. In all applications it is essential to match sources and sensors spectrally to maximize energy transfer. For diminished photon sources, the photomultiplier excels owing to a photoemissive cathode followed by high multiplicative gain to 10^6 from its cascaded dynode structure. The high gain and inherent low noise provided by coordinated multiplication results in a widely applicable sensor, except for the infrared region. Presently, the photomultiplier vacuum electron ballistic structure does not have a solid-state equivalent.

The measurement of infrared radiation is difficult because of the low energy of the infrared photon. This sensitivity deficiency has been overcome by thermally responsive resistive bolometer microsensors employing high-T_c superconductive films, whereby operation is maintained near the film transition temperature such that small temperature variations from infrared photons provide large resistance changes with gains to 10^3. Microsensor fabrication enhances reliability, and extension to arraying of elements is described by Figure 1.15, with image intensity, $\bar{I} = (x, y)$, quantized into a gray-level representation, $f(\bar{I})$. This versatile imaging device is employed in applications ranging from analytical spectroscopy to night vision and space defense.

Table 1.4 Sensor Relative Performance

Device	λ Region	$I_{photocurrent}/F_{photons/sec}$	Application
Photo diode	UV–near IR	10^{-3} amp/watt	Photonic detector
Photoconductive	Visible–near IR	1 amp/watt	Photo controller
Bolometer	Near IR–far IR	10^3 amps/watt	Superconducting IR
Photomultiplier	UV–near IR	10^6 amps/watt	Spectroscopy

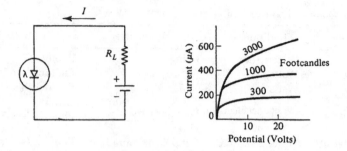

Figure 1.13 Photo diode cell characteristics.

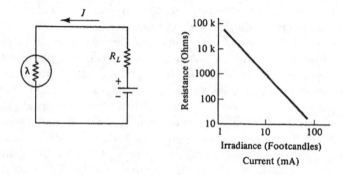

Figure 1.14 Photoconductive characteristics.

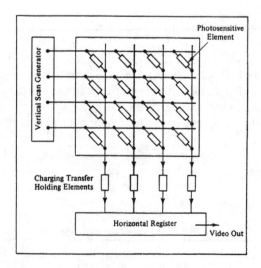

Figure 1.15 Quantum sensor array.

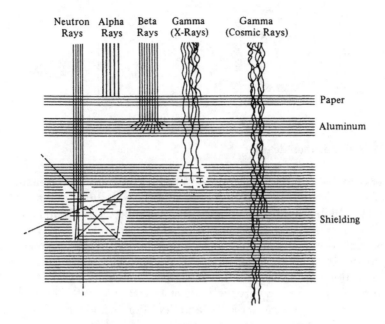

Figure 1.16 Nuclear radiation characteristics.

A property common to all nuclear radiation is its ability to interact with the atoms that constitute all matter. The nature of the interaction varies with the different components of radiation, as illustrated in Figure 1.16. These components are responsible for interactions with matter that generally produce ionization of the medium through which they pass. This ionization is the principal effect used to detect the presence of nuclear radiation. Alpha and beta rays often are not encountered because of their attenuation. Instruments to detect nuclear radiation, therefore, most commonly are constructed to measure gamma radiation and its scintillation or luminescent effect. The ionization rate in Roentgens per hour is a preferred measurement unit for gamma-ray photons in electron volts (eV), produced by radioactive element decay from higher to lower energy states with a disintegration rate in Curies (Ci). A distinction also should be made between disintegrations in counts per minute and ionization rate. The count-rate measurement is useful for half-life determination and nuclear detection but does not provide exposure rate information for interpretation of degree of hazard. The estimated yearly radiation dose to persons in the United States is 0.25 Roentgen (R). A high radiation area is defined where radiation levels exceed 0.1 R per hour, and requires posting of a caution sign.

Methods for detecting nuclear radiation are based on means for measuring the ionizing effects of these radiations. Mechanizations fall into two categories: pulse-type detectors of ionizing events and ionization-current detectors that employ an ionization chamber to provide an averaged radiation effect. The first category includes Geiger-Mueller tubes and more sensitive scintillation

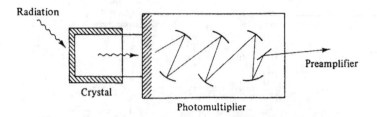

Figure 1.17 Scintillation detector.

counters capable of individual counts. Detecting the individual ionizing scintillations is aided by an activated crystal such as sodium iodide optically coupled to a high-amplification photomultiplier tube, as shown in Figure 1.17. Ionization-current detectors primarily are employed in health-physics applications, such as industrial areas subject to high radiation levels. An ion chamber is followed by an amplifier whose output is calibrated in Roentgens per hour ionization rate. This method is necessary where pulse-type detectors are inappropriate because of a very high rate of ionization events. Practical industrial applications of nuclear radiation and detection include thickness gauges, nondestructive testing such as x-ray inspection, and chemical analysis such as by neutron activation.

1.5 Analytical sensors

The relationship between process and analytical measurements often is only a matter of sensor location; process measurements are acquired during real-time physical events whereas analytical measurements may be acquired post-event offline as an *ex situ* assay. These measurements provide useful describing functions that increasingly are applied as a substitute for conventional mathematical process models. Chemical sensors are employed to determine the presence, concentration, or quantity of elemental or molecular analytes. These sensors may be divided into two classes: electromagnetic, involving filtered optical and atomic mass unit spectroscopy; or electrochemical, involving the selectivity of charged species. Quantum spectroscopy, described by Figure 1.18, offers specific chemical measurements utilizing wavelength-selective filters from UV to near IR coupled to a photoemissive

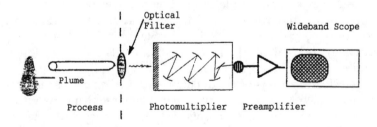

Figure 1.18 Optical spectrometer structure.

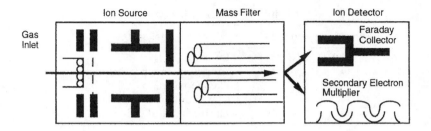

Figure 1.19 Mass spectrometer sensor.

photomultiplier with output displayed by a wideband oscilloscope. Alternatively, mass spectrometry chemical analysis is performed at high vacuum, typically employing a quadrupole filter, shown in Figure 1.19, with sample gas energized by an ion source. The mass filter selects ions determining specific charge-to-mass ratios, employing both electric and magnetic fields with the relationship $mV^2 = 2eV$, which are subsequently collected for gas samples of interest to express their chemical identity by an ion detector whose current intensity is displayed versus atomic mass unit (AMU).

Online measurements of industrial processes and chemical streams often require the use of selective chemical analyzers to control a processing unit. Examples include oxygen for boiler control, sulfur oxide emissions from combustion processes, and hydrocarbons associated with petroleum refining. Laboratory instruments such as gas chromatographs generally are not used for online measurements, primarily because they analyze all compounds present simultaneously rather than a single one of interest.

The dispersive infrared analyzer is the most widely used chemical analyzer, owing to the range of compounds it can be configured to measure. Operation is by the differential absorption of infrared energy in a sample stream in comparison to that of a reference cell. Measurement is by deflection of a diaphragm separating the sample and reference cells, which in turn detunes an oscillator circuit to provide an electrical analog of compound concentration. Oxygen analyzers usually are of the amperometric type in which oxygen is chemically reduced at a gold cathode, resulting in a current flow from a silver anode as a function of this reduction and oxygen concentration. In a paramagnetic wind device, a wind effect is generated when a mixture containing oxygen produces a gradient in a magnetic field. Measurement is derived by the thermal cooling effect on a heated resistance element, forming a thermal anemometer. Table 1.5 presents basic electrochemical analyzer methods, and Figure 1.20 a basic gas analyzer system with calibration.

Also in this group are pH, conductivity, and ion-selective electrodes. pH defines the balance between the hydrogen ions, H^+, of an acid and the hydroxyl ions, OH^-, of an alkali, in which one type can be increased only at the expense of another. A pH probe is sensitive to the presence of H^+ ions in solution, thereby representing the acidity or alkalinity of a sample. All of these ion-selective electrodes are based on the Nernst equation (1.7), which

Table 1.5 Chemical Analyzer Methods

Compound	Analyzer
CO, SO_x, NH_x	Infrared
O	Amperometric, paramagnetic
HC	Flame ionization
NO_x	Chemiluminescent
H_2S	Electrochemical cell

typically provides a 60-mV potential change for each tenfold change in the activity of a monovalent ion.

$$V_o = V + \frac{F}{n} \log(ac + s_1 a_1 c_1 + \ldots) \text{ volts} \tag{1.7}$$

where

V_o = voltage between sensing and reference electrodes

V = electrode base potential

F = Nernst factor, 60 mV at 25°C

n = ionic charge, 1 monovalent, 2 bivalent, etc.

a = ionic activity

c = concentration

s = electrode sensitivity to interfering ions

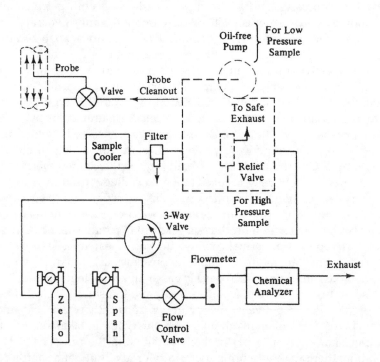

Figure 1.20 Calibrated gas analyzer.

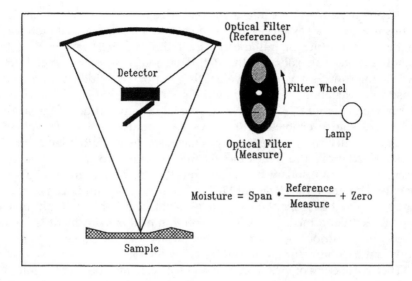

Figure 1.21 Near-infrared analytical sensor.

Online near-infrared sensors are typically employed for process and quality measurements, primarily for the control of content and film thickness of water and organic materials contained by products from paper to polymers to steel. Measurement error to 0.1%FS is available with sample illumination, absorption, and reflectance at near-infrared wavelengths between 1100 and 2500 nm. Figure 1.21 illustrates a mechanization whereby different infrared filters are used for measurement (absorption wavelength) and reference (reflectance wavelength) to achieve a ratio that compensates for variabilities encountered in the application environment. Sensor calibration is instituted by means of span and zero that, respectively, adjust sensitivity and offset.

The concept of computer-based instruments arose with the advent of inexpensive computation, furthered by the personal computer, which permitted networking discrete instruments into sophisticated automated test systems beginning in the 1970s. The evolution of more efficient data acquisition and presentation, resulting from user-defined programmability and reconfigurability, continues through the present to provide a more computationally intensive instrumentation framework. Contemporary virtual instruments accordingly are capable of elevating fundamental sensor data to a substantially higher attribution, enabling more complex cognitive interpretation. Multifunction I/O hardware is typically combined with application development software on a personal computer platform for the realization of specific virtual instruments, such as the following microwave microscopy example for sample assays in manufacturing and biomedical applications.

A benefit of microwave microscopy is micron-resolution sample imaging of subsurface as well as surface properties. With microwave excitation wavelengths on the order of millimeters, their limiting half-wavelength Abbe resolution barrier is extended through detection of shorter near-field evanescent microwave spatial wavelength components aided by enhanced transducer excitation, sensitivity, and signal processing. With the virtual instrument of Figure 1.22, a 100-Hz sinusoidal 1F signal dither of the 30-GHz microwave source enables synthesis of 2F spatial frequency components associated with a sample. Sensitivity is increased by an automatic controller whose DC-controlled variable tunes the microwave source, as does the dither signal, to obtain a resonant frequency shift that maximizes the spatial components. The greater a 2F signal DFT-normalized magnitude achieved, the more selective the resonant curve quality factor and frequency shift images become relative to the sample. The resolution of this instrument is limited by –55 dB of spurious noise from contemporary gigahertz sources, which is equivalent to 9-bit accuracy according to Table 5-7.

The architecture of virtual instrument software may be divided into two layers: (1) measurement and configuration services and (2) application development tools. Measurement and configuration services contain prescriptive software drivers for interfacing hardware I/O devices as subroutines accessed by graphical icons. Configuration utilities are included for naming and setting hardware channel attributes such as amplitude scaling. Software selected for application development may be sourced separately from hardware devices only when compatibility is ensured.

Graphical programs typically consist of an icon diagram including a front panel that serves as the source code for an application program. The front panel provides a graphical user interface for functions to be executed concurrently. The LabVIEW diagram of Figure 1.23 shows a View Image module for generating the data display images shown in Figure 1.22, which may be exported as Matlab files. When this program is initiated, the front panel, enumerated (1), defines display visibility attributes. Assets within a 'while loop' are then executed cyclically until control Done is set false, enumerated (2), allowing conditional expressions to break this 'while loop.' The metronome icon describes a 50-msec interval within which the 'while loop' iterates. The data structure that performs this image generation executes sequentially.

The concentric window enumerated (3) is analogous to a 'case' statement in the C language controlled by the Boolean Load variable. The 'case' procedure occurs when Load is true. The icons within (3) allow a user to locate an image file, whereas the icons within (4) provide a subprogram for extracting microscopy image content for arrays representing frequency shift and quality factor data. A 'sequence' data structure performs total image formatting. Icons within (4) select strings contained in (5) to write data images as XY vectors. Note that this code is set to display X and Y in millimeter units.

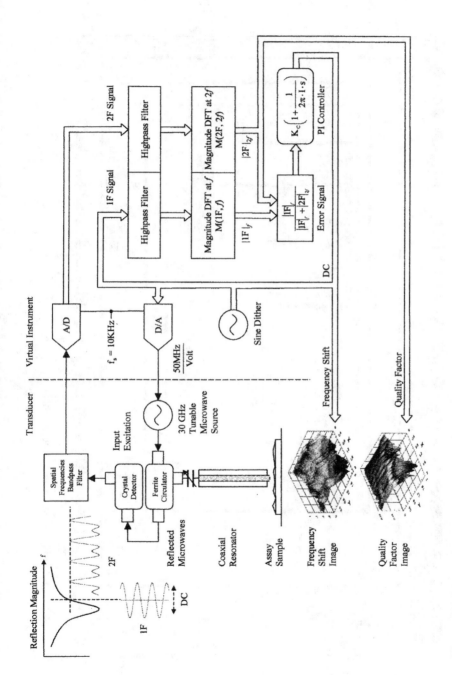

Figure 1.22 Microwave microscopy biomedical instrument.

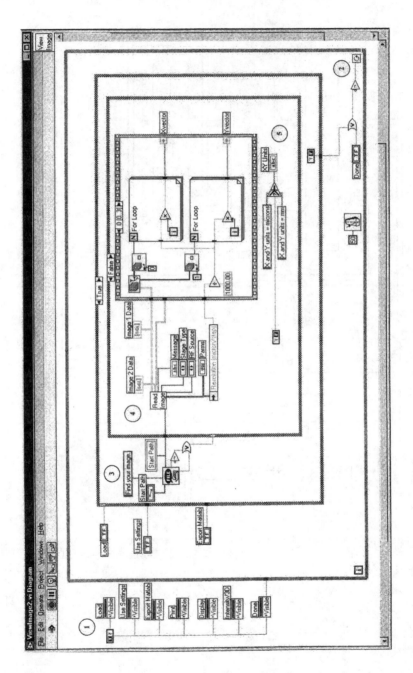

Figure 1.23 LabVIEW display generation graphical program.

Bibliography

1. Gardner, J.W., *Microsensors*, John Wiley & Sons, New York, 1994.
2. Garrett, P.H. et al., *Advanced Instrumentation and Computer I/O Design*, IEEE Press, New York, 1994).
3. Garrett, P.H. et al., "Emerging Methods for the Intelligent Processing of Materials," *Journal of Materials Engineering and Performance*, 2(5), October 1993 p. 727.
4. Garrett, P.H. (contributing author), *Handbook of Industrial Automation*, eds. Shell, R.L. and Hall, E.L. Marcel Dekker, New York, 2000.
5. Garrett, P.H. et al., "Self-Directed Processing of Materials," *IFAC, Engineering Applications of Artificial Intelligence*, 12, August 1999 p. 479.
6. Jackson, A.G. et al. "Sensor Principles and Methods for Measuring Physical Properties," *Journal of Materials*, 48(9), September 1996, p. 16.
7. Kovacs, G.T.A., *Micromachined Transducers Sourcebook*, McGraw-Hill, New York, 1998.
8. Malas, J. et al., "Emerging Sensors for the Intelligent Processing of Materials," *Journal of Materials*, 48(9), September 1996 p. 16.
9. Norton, H.N., *Handbook of Transducers for Electronic Measuring Systems*, Prentice Hall, New York, 1969.
10. Petriu, E.M., Ed., *Instrumentation and Measurement Technology and Applications*, IEEE Technology Update Series, 1998.
11. Prensky, S.D. and Castellucis, R.L., *Electronic Instrumentation*, 3rd ed., Prentice Hall, New York, 1982.
12. Rabinovich, S.G., *Measurement Errors and Uncertainties*, 2nd ed., Springer-Verlag, Heidelberg, 1999.
13. Tabib-Azar, M. et al., "Near Real-Time Monitoring of Thin-Film Materials and Their Interfaces Using Evanescent Microwave Probes," AFRL-ML-WP-TR-1999-4074, April 1999, Wright-Patterson AFB, OH 45433.

chapter two

Instrumentation amplifiers and parameter errors

2.0 Introduction

This chapter deals with the devices and circuits that comprise the electronic amplifiers of linear systems used in instrumentation applications. The discussion begins with the temperature limitations of semiconductor devices and is extended to differential amplifiers and an analysis of their parameters for understanding operational amplifiers from the perspective of their internal stages. This includes gain–bandwidth–phase stability relationships and interactions in multiple amplifier systems. An understanding of the capabilities and limitations of operational amplifiers is essential as a preface to instrumentation amplifiers.

An instrumentation amplifier usually is the first electronic device encountered in a signal acquisition system, and in large part it is responsible for the ultimate data accuracy attainable. Present instrumentation amplifiers possess sufficient linearity, common-mode rejection ratio (CMRR), low noise, and precision for total errors in the microvolt range. Five categories of instrumentation amplifier applications are described with representative contemporary devices and parameters provided for each. These parameters are then used to compare amplifier circuits for implementations ranging from low input voltage error to wide bandwidth applications.

2.1 Device temperature characteristics

The elemental semiconductor device in electronic circuits is the pn junction; among its forms are diodes, bipolar, and FET (field effect) transistors. The availability of free carriers that result in current flow in a semiconductor is a direct function of the applied thermal energy. At room temperature, taken as 20°C (293 K, above absolute zero), there is abundant energy to liberate the valence electrons of a semiconductor. These carriers are then free to drift under the influence of an applied potential. The magnitude of

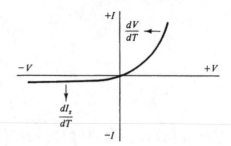

Figure 2.1 pn junction temperature dependence.

this current flow is essentially a function of the thermal energy instead of the applied voltage and accounts for the negative temperature coefficient exhibited by semiconductor devices (increasing current with increasing temperature).

The primary variation associated with reverse-biased pn junctions is the change in reverse saturation current, I_s, with temperature. I_s is determined by device geometry and doping, with a variation of 8% per degree centigrade both in silicon and germanium, doubling every 10°C rise. This behavior is shown by Figure 2.1 and equation (2.1). Forward-biased pn junctions exhibit a decreasing junction potential, having an expected value of –2.0 mV per degree centigrade rise, as defined by equation (2.2). The $\frac{dV}{dT}$ temperature variation is shown to be the difference between the forward junction potential, V, and the temperature of I_s. This relationship is the source of the voltage offset drift with temperature exhibited by semiconductor devices. The volt equivalent of temperature is an empirical model in both equations defined as $V_T = (273\ \mathrm{K} + T°\mathrm{C})/11{,}600$, having an expected value of 25 mV at room temperature.

$$\frac{dI_s}{dT} = I_s \bullet \frac{d(\ln I_s)}{dT}\ \mathrm{A/°C} \tag{2.1}$$

$$\frac{dV}{dT} = \left(\frac{V}{T} - \frac{V_T}{I_s} \bullet \frac{dI_s}{dT} \right)\ \mathrm{V/°C} \tag{2.2}$$

2.2 Differential amplifiers

The first electronic circuit encountered by a sensor signal in a data acquisition system typically is the differential input stage of an instrumentation amplifier. The balanced bipolar differential amplifier of Figure 2.2(a) is an important circuit used in many linear applications. Operation with symmetrical ± power supplies as shown results in the input base terminals being at 0 V under quiescent conditions. Due to the interaction that occurs in this emitter-coupled circuit, the algebraic difference signal applied across the input terminals is the effective drive signal, whereas equally applied input signals

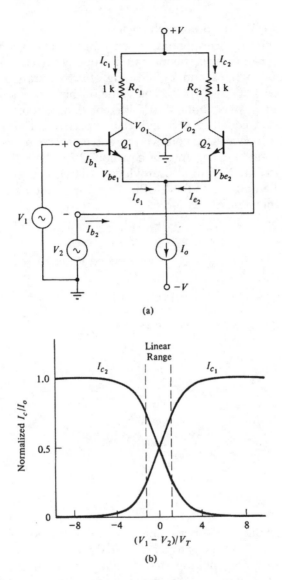

(a)

(b)

Figure 2.2 Differential DC amplifier and normalized transfer curves; $h_{fe} = 100$, $h_{ie} = 1k$, $h_{oe} = 10^{-6}$ mho.

are cancelled by the symmetry of the circuit. With reference to a single-ended output, V_{o_2}, amplifier Q_1 may be considered an emitter follower with the constant-current source an emitter load impedance in the megohm range. This results in a noninverting voltage gain for Q_1 very close to unity (0.99999) that is emitter coupled to the common-emitter amplifer, Q_2, where Q_2 provides the differential voltage gain $A_{v_{diff}}$ by equation (2.3).

Differential amplifier volt-ampere transfer curves are defined by Figure 2.2(b), where the abscissa represents normalized differential input voltage $(V_1 - V_2)/V_T$. The transfer characteristics are shown to be linear about the operating point, corresponding to an input-voltage swing of approximately 50 mV (± 1 V_T unit). The maximum slope of the curves occurs at the operating point of $I_o/2$ and defines the effective transconductance of the circuit as $\Delta I_C/\Delta(V_1 - V_2)/V_T$. The value of this slope is determined by the total current, I_O, of equation (2.4). Differential input impedances, R_{diff} and R_{cm}, are defined by equations (2.5) and (2.6), respectively. The effective voltage gain cancellation between the noninverting and inverting inputs is represented by the common-mode gain, Av_{cm} of equation (2.7). The ratio of differential gain to common-mode gain also provides a dimensionless figure of merit for differential amplifiers as the CMRR. This is expressed by equation (2.8), having a typical value of 10^5.

$$\text{Av}_{diff} = \frac{h_{fe}R_c}{2h_{ie}} \text{ (single-ended } V_{o_2}) = 50 \tag{2.3}$$

$$I_O = I_{s1} \bullet \exp{(V_{be1}/V_T)} + I_{s2} \bullet \exp{(V_{be2}/V_T)} = 1 \text{ mA} \tag{2.4}$$

$$R_{diff} = \frac{4V_T h_{fe}}{I_O} = 10 \text{ K}\Omega \tag{2.5}$$

$$R_{cm} = \frac{h_{fe}}{h_{oe}} = 100 \text{ M}\Omega \tag{2.6}$$

$$\text{Av}_{cm} = \frac{h_{oe}R_c}{2} = 5 \times 10^{-4} \tag{2.7}$$

$$\text{CMRR} = \frac{\text{Av}_{diff}}{\text{Av}_{cm}} = 10^5 \tag{2.8}$$

The performance of operational and instrumentation amplifiers is largely determined by the errors associated with their input stages. It is convention to express these errors as voltage and current offset values, including their variation with temperature with respect to the input terminals, so that various amplifiers may be compared on the same basis. In this manner, factors such as the choice of gain and the amplification of the error values do not result in confusion concerning their true magnitude. It is also notable that the symmetry provided by the differential amplifier circuit primarily serves to offer excellent DC stability and the minimization of input errors in comparison with those of nondifferential amplifiers.

The base-emitter voltages of a group of the same type of bipolar transistors at the same collector current are typically only within 20 mV. Operation of a differential pair with a constant-current emitter sink as

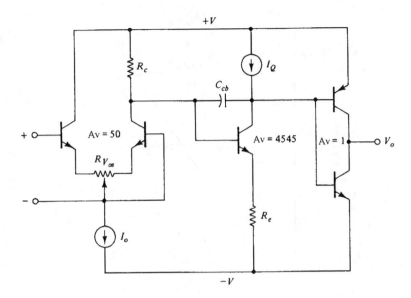

Figure 2.3 Elemental operational amplifier circuit.

shown in Figure 2.2(a), however, provides a V_{be} match of V_{os} of about 1 mV. Equation (2.9) defines this input offset voltage and its dependence on the mismatch in reverse saturation current I_s between the differential pair. This mismatch results from variations in doping and geometry of the devices during their manufacture. Offset adjustment is frequently provided by the introduction of an external trimpot, $R_{V_{os}}$, in the emitter circuit, shown in Figure 2.3. That permits the incremental addition and subtraction of emitter voltage drops to zero V_{os} without disturbing the emitter current I_o.

Of greater concern is the offset voltage drift with temperature, dV_{os}/dT. This input error results from mistracking of V_{be1} and V_{be2}, described by equation (2.10), and is difficult to compensate. However, the differential circuit reduces dV_{os}/dT to 2 µV/°C from the –2 mV/°C for a single device of equation (2.2), for an improvement factor of 1/1000. By way of comparison, JFET differential circuit V_{os} is larger and on the order of 10 mV, and dV_{os}/dT is typically 5 µV/°C. Minimization of these errors is achieved by matching the device pinch-off voltage parameter. Bipolar input bias current offset and offset current drift are described by equations (2.11) and (2.12), respectively, and have their genesis in a mismatch in current gain ($h_{fe1} \neq h_{fe2}$). JFET devices intrinsically offer lower input bias currents and offset current errors in differential circuits, which is advantageous for the amplification of current-type sensor signals. However, the rate of increase of JFET bias current with temperature is exponential, as illustrated in Figure 2.4, and results in values that exceed bipolar input bias currents at temperatures

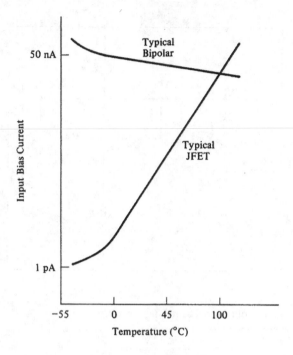

Figure 2.4 Transistor input current temperature drift.

beyond 100°C, thereby limiting the utility of JFET differential amplifiers above this temperature.

$$V_{os} = V_T \ln \frac{I_{s2}}{I_{s1}} \bullet \frac{I_{e1}}{I_{e2}} = 1 \text{ mV} \tag{2.9}$$

$$\frac{dV_{os}}{dT} = \frac{dV_{be1}}{dT} - \frac{dV_{be2}}{dT} = 2 \text{ μV/°C} \tag{2.10}$$

$$I_{os} = I_{b1} - I_{b2} = 50 \text{ nA} \tag{2.11}$$

$$\frac{dI_{os}}{dT} = B \bullet I_{os} = 0.25 \text{ nA/°C} \tag{2.12}$$

$$B = -0.005/°C > 25°C$$

2.3   Operational amplifiers

Most operational amplifiers are of similar design, as described by Figure 2.4, and consist of a differential-input stage cascaded with a high-gain interstage followed by a power-output stage. Operational amplifiers are characterized by very high gain at DC and a uniform rolloff in this gain with frequency. This enables these devices to accept feedback from arbitrary networks with high stability and simultaneous DC and AC amplification. Consequently, such networks can accurately impart their characteristics to electronic systems with

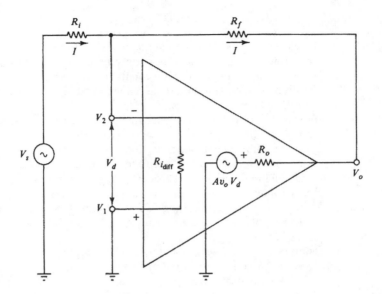

Figure 2.5 Inverting operational amplifier. Since $R_{i_{diff}} \to \infty, V_d = \frac{V_o}{Av_o} \to o$ as $|Av_o| \to \infty$.

negligible degradation. The earliest integrated-circuit amplifier was offered in 1963 by Texas Instruments, but the Fairchild 709 introduced in 1965 was the first operational amplifier to achieve widespread application. Improvements in design resulted in second-generation devices such as the National LM108. Advances in fabrication technology enabled amplifiers such as the Analog Devices OP-07 with improved performance overall. Subsequent refinements are represented by devices including the Linear LTC-1250, featuring zero drift and ultra-low noise. It is notable that contemporary operational amplifier circuits are structured around a high-gain innerstage employing a constant-current source active load. The gain stage active load impedance of approximately 500 kΩ ratioed with an emitter resistance, R_e, approximating 100 Ω, shown in Figure 2.4, is responsible for high overall Av_o.

$$Av_c = \frac{V_o}{V_s} = \frac{-IR_f}{IR_i} = \frac{-R_f}{R_i} \qquad (2.13)$$

The circuit for an inverting operational amplifier is shown in Figure 2.5. The cascaded innerstage gains of Figure 2.4 provide a total open-loop gain, Av_o, of 227,500, enabling realization of the ideal closed-loop gain, Av_c, representation of equation (2.13). In practice, the Av_o value cannot be utilized without feedback because of nonlinearities and instability. The introduction of negative feedback between the output and inverting input also results in a virtual ground with equilibrium current conditions maintaining $V_d = V_1 - V_2$ at zero. Classification of operational amplifiers is primarily determined by the active devices that implement the amplifier differential input. Table 2.1 delineates this classification.

Table 2.1 Operational and Instrumentation Amplifiers

Bipolar	Prevalent type used for a wide range of signal-processing applications. Good balance of performance characteristics.
FET	Very high input impedance. Frequently employed as an instrumentation-amplifier preamplifier. Exhibits larger input errors than bipolar devices.
CAZ*	Bipolar device with auto-zero circuitry for internally measuring and correcting input error voltages. Provides low-input-uncertainty amplification.
BiFET[†]	Combined bipolar and FET circuit for extended performance. Intended to displace bipolar devices in general-purpose applications.
Superbeta	A bipolar device approaching FET input impedance with the lower bipolar errors. A disadvantage is lack of device ruggedness.
Micropower	High-performance operation down to 1-V supply powered from residual system potentials. Employs complicated low-power circuit equivalents for implementation.
Isolation	An internal barrier device using modulation or optical methods for very high isolation. Medical and industrial applications.
Chopper	DC-AC-DC circuit with a capacitor-coupled internal amplifier providing very low offset errors for minimum input uncertainty.
Varactor	Varactor diode input device with very low input bias currents for current-amplification applications such as photomultipliers.
Vibrating capacitor	A special input circuit arrangement requiring ultra-low input bias currents for applications such as electrometers.

*Commutating Auto Zero
[†]Bipolar FET

According to negative-feedback theory, an inverting amplifier will be unstable if its gain is equal to or greater than unity when the phase shift reaches $-180°$ through the amplifier. This is so because an output-to-input relationship will also have been established providing an additional $-180°$ by the feedback network. The relationships between amplifier gain, bandwidth, and phase are described by Figure 2.6 and equations (2.14) through (2.16) for an example closed-loop gain, Av_c, value of 100. Each discrete innerstage contributes a total of $-90°$ to the cumulative phase shift, ϕ_t, with $-45°$ realized at the respective -3-dB frequencies. The high-gain stage -3-dB frequency of 10 Hz is attributable to the dominant-pole compensating capacitance, C_{cb}, shown in Figure 2.3.

$$A_{V_o} = \frac{227,250}{\left(1 + j\dfrac{f}{10\ \text{Hz}}\right)\left(1 + j\dfrac{f}{1\ \text{MHz}}\right)\left(1 + j\dfrac{f}{25\ \text{MHz}}\right)} \qquad (2.14)$$

$$\phi_t = -\tan^{-1}\left(\frac{f}{10\ \text{Hz}}\right) - \tan^{-1}\left(\frac{f}{1\ \text{MHz}}\right) - \tan^{-1}\left(\frac{f}{25\ \text{MHz}}\right) \qquad (2.15)$$

$$\text{Phase margin} = 180° - \phi_t \qquad (2.16)$$

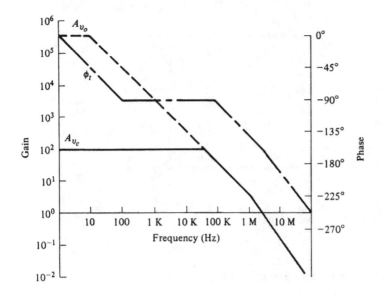

Figure 2.6 Operational amplifier gain–bandwidth–phase relationships.

2.4   Instrumentation amplifiers

The acquisition of accurate measurement signals, especially low-level signals in the presence of interference, requires amplifier performance beyond the typical signal conditioning capabilities of operational amplifiers. An instrumentation amplifier is usually the first electronic device encountered by a sensor in a signal acquisition channel, and in large part it is responsible for the ultimate data accuracy attainable. Present instrumentation amplifiers possess sufficient linearity, stability, and low noise for total error in the microvolt range even when subjected to temperature variations, and is equivalent to the nominal thermocouple effects exhibited by input lead connections. High CMRR is essential for achieving the amplifier performance of interest with regard to interference rejection, and for establishing a signal ground reference at the amplifier that can accommodate the presence of ground-return potential differences. High amplifier input impedance is also necessary to preclude input signal loading and voltage divider effects from finite source impedances, and to accommodate source-impedance imbalances without degrading CMRR. The precision gain values possible with instrumentation amplifiers, such as 1000.000, are equally important to obtain accurate scaling and registration of measurement signals.

The relationship of CMRR to the output signal, V_o, for an operational or instrumentation amplifier is described by equation (2.17) and is based on the derivation of CMRR provided by equation (2.8). For the operational amplifier subtractor circuit of Figure 2.7, Av_{diff} is determined by the feedback-to-input

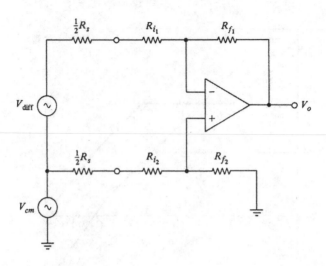

Figure 2.7 Subtractor instrumentation amplifier.

resistor ratios (R_f/R_i) with practically realizable values to 100, and Av_{cm} is determined by the mismatch between feedback and input resistor values attributable to their tolerances. Consequently, the Av_{diff} for a subtractor circuit may be obtained from equation (2.18) and as tabulated in Table 2.2 to determine the expected CMRR value for specified resistor tolerances. Notice that CMRR increases with Av_{diff} by the numerator of equation (2.8), but Av_{cm} is constant because of its normalization by the resistor tolerance chosen.

$$V_o = Av_{diff} \bullet V_{diff} + Av_{cm} \bullet V_{cm}$$

$$= Av_{diff} \bullet V_{diff}\left(1 + \frac{1}{CMRR} \bullet \frac{V_{cm}}{V_{diff}}\right) \tag{2.17}$$

$$CMRR_{subtractor} = \frac{\dfrac{1}{2}\left(\left|\dfrac{R_{f2} \pm \Delta R_{f2}}{R_{i2} \pm \Delta R_{i2}}\right| + \left|\dfrac{R_{f1} \pm \Delta R_{f1}}{R_{i1} \pm \Delta R_{i1}}\right|\right)}{\left|\dfrac{R_{f2} \pm \Delta R_{f2}}{R_{i2} \pm \Delta R_{i2}}\right| - \left|\dfrac{R_{f1} \pm \Delta R_{f1}}{R_{i1} \pm \Delta R_{i1}}\right|} \tag{2.18}$$

Table 2.2 Subtractor CMRR Expected Values

Resistor Tolerance	5%	2%	1%	0.1%
$Av_{cm\ subtractor}$	0.1	0.04	0.02	0.002
$CMRR_{subtractor}\ (xAv_{diff})$	10	25	50	500

The subtractor circuit is capable of typical values of CMRR to 10^4, and its implementation is economical owing to the requirement for a single operational amplifier. However, its specifications are usually marginal when compared with the requirements of typical signal acquisition applications. For example, each implementation requires the matching of four resistors, and the input impedance is constrained to the value of R_i chosen. For modern bipolar amplifiers, such as the Analog Devices OP-07 and Burr Brown OPA-128 devices with gigohm internal resistances, megohm R_i values are allowable to prevent input voltage divider effects resulting from an imbalanced kilohm R_s source resistance. Low-bias-current amplifiers are essential for current sensors including nuclear gauges, pH probes, and photomultiplier tubes. The OPA-128 offers a balance of input parameters for this application with an I_{os} of 30 fA and typical current sensor R_s values of 10 MΩ. The compensating resistor, R_c, shown in Figure 2.8 is matched to R_s in order to preserve CMRR. The five amplifiers presented in Table 2.3 beneficially permit the comparison of parameters that influence performance in specific amplifier applications; the CMRR entries described are expected in-circuit values.

The three-amplifier instrumentation amplifier of Figure 2.9, exampled by the AD624, offers improved performance overall compared to the foregoing subtractor circuit, with in-circuit CMRR_{3ampl} values of 10^5 and the absence of problematic external discrete input resistors. To minimize output noise and offsets with this amplifier its subtractor Av_{diff} is normally set to unity gain. The first stage of this amplifier also has a unity Av_{cm}, owing to its differential-input-to-differential-output connection, which results in identical first-stage CMRR and Av_{diff} values. Amplifier internal resistance trimming consequently achieves the nominal subtractor Av_{cm} value shown in equation (2.19).

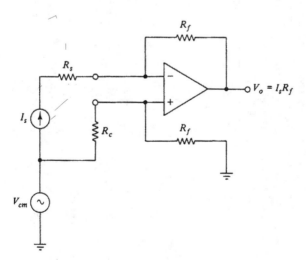

Figure 2.8 Differential current-voltage amplifier.

Table 2.3　Amplifier Input Parameters Define Interface Applications

Symbol	Low-Offset Voltage OP-07	Low-Bias Current OPA-128	Three-Amplifier Instrumentation AD624	High-Voltage Isolation AD215	Wideband Video OPA-646	Comment
V_{OS}	10 µV	140 µV	25 µV	0.4 mV	1 mV	Offset voltage
$\dfrac{dV_{OS}}{dT}$	0.2 µV/°C	5 µV/°C	0.25 µV/°C	2 µV/°C	12 µV/°C	Offset voltage drift
I_{OS}	0.3 nA	30 fA	10 nA	300 nA	0.4 µA	Offset current
$\dfrac{dI_{OS}}{dT}$	5 pA/°C	Negligible	20 pA/°C	1 nA/°C	10 nA/°C	Offset current drift
Sr	0.3 V/µsec	3 V/µsec	5 V/µsec	6 V/µsec	180 V/µsec	Slew rate
f_{hi}	600 KHz	500 KHz	1 MHz	120 KHz	650 MHz	Unity gain bandwidth
CMRR	10^4	10^4	10^5	10^5	10^4	Av_{diff}/Av_{cm}
V_{cm}	10 Vrms	10 Vrms	10 Vrms	1500 Vrms	10^4	Maximum applied volts
V_n rms	10 nV/$\sqrt{Hz}$	27 nV/$\sqrt{Hz}$	4 nV/$\sqrt{Hz}$	Negligible	7.5 nV/$\sqrt{Hz}$	Voltage noise
$f(Av)$	0.01%	0.01%	0.001%	0.0005%	0.025%	Gain nonlinearity
$\dfrac{dAv}{dT}$	50 ppm/°C	50 ppm/°C	5 ppm/°C	15 ppm/°C	50 ppm/°C	Gain drift
R_{diff}	$8 \times 10^7\ \Omega$	$10^{13}\ \Omega$	$10^9\ \Omega$	$10^{12}\ \Omega$	15 KΩ	Differential resistance
R_{cm}	$2 \times 10^{11}\ \Omega$	$10^{13}\ \Omega$	$10^9\ \Omega$	$5 \times 10^9\ \Omega$	1.6 MΩ	Common-mode resistance

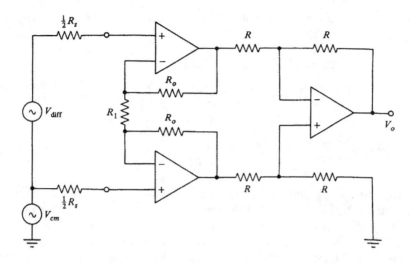

Figure 2.9 Three-amplifier instrumentation amplifier.

The differential output instrumentation amplifier, illustrated by Figure 2.10, offers increased common-mode rejection via equation (2.20) over the three-amplifier circuit from the addition of a second output subtractor. By comparison, a single subtractor permits a full-scale 24-V_{p-p} output signal swing whereas dual subtractors deliver a full-scale 48-V_{p-p} output signal from opposite polarity swings of the ±15-V DC power supplies for each signal half cycle. The effective output gain doubling combined with first-stage gain provides $CMRR_{diff}$ output values to 10^6. This advanced amplifier circuit permits high-performance analog signal acquisition and the continuation of common-mode interference rejection over a signal transmission channel,

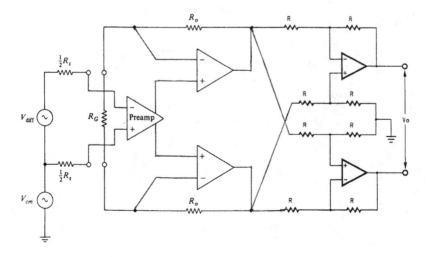

Figure 2.10 Differential output instrumentation amplifier.

with termination by a remote differential-to-single-ended subtractor amplifier.

$$\text{CMRR}_{3\,\text{ampl}} = \text{CMRR}_{1\text{st stage}} \bullet \text{CMRR}_{\text{subtractor}}$$

$$= \frac{\text{Av}_{\textit{diff}\,1\text{st stage}}}{1} \bullet \frac{1}{\text{Av}_{cm\,\text{subtractor}}} = \left(1 + \frac{2R_0}{R_1}\right) \bullet \left(\frac{1}{0.001}\right) \qquad (2.19)$$

$$\text{CMRR}_{\text{diff output}} = \left(1 + \frac{2R_0}{R_G}\right) \bullet \left(\frac{2}{0.001}\right) \qquad\qquad (2.20)$$

Isolation amplifiers are advantageous for very noisy and high-voltage environments plus the interruption of ground loops. They further provide galvanic isolation typically on the order of 1 µA input-to-output leakage. The front end of the iso amplifier is similar to an instrumentation amplifier, as shown in Figure 2.11, and is operated from an internal DC-DC isolated power supply to ensure isolation integrity and for external sensor excitation purposes. As a consequence, these amplifiers do not require sourcing or sinking external bias current and function normally with fully floating

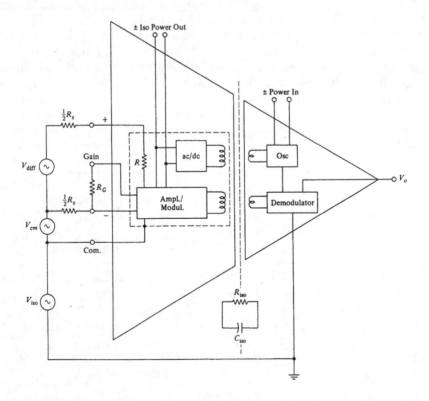

Figure 2.11 Isolation instrumentation amplifier.

sensors. Most designs also include a 100-KΩ series input resistor, R, to limit catastrophic fault currents. Typical isolation barriers have an equivalent circuit of 10^{11} Ω shunted by 10 pF representing R_{iso} and C_{iso}. An input-to-output V_{iso} rating of 1500 V_{rms} is common and has a corollary isolation-mode rejection ratio (IMRR) with reference to the output. CMRR values of 10^5 relative to the input common and IMRR values to 10^8 with reference to the output are available at 60 Hz. This capability makes possible the accommodation of two sources of interference, V_{cm} and V_{iso}, both frequently encountered in sensor applications. The performance of this connection is described by equation (2.21).

$$V_o = \mathrm{Av}_{diff} \bullet V_{diff} \left(1 + \frac{1}{\mathrm{CMRR}} \bullet \frac{V_{cm}}{V_{diff}} \right) + \frac{V_{iso}}{\mathrm{IMRR}} \qquad (2.21)$$

High-speed data conversion and signal conditioning circuits capable of accommodating pulse and video signals require wideband operational amplifiers. Such amplifiers are characterized by their settling time, delay, slew rate, and transient subsidence, described in Figure 2.12. Parasitic reactive circuit elements and carelessly planned circuit layouts can result in performance derogation. Amplifier slew rate depends directly on the product of the output voltage amplitude and signal frequency, and this product cannot exceed the slew rate specification of an amplifier if linear performance is to be realized. For example, a 1-V_{pp} sinewave signal at a frequency of

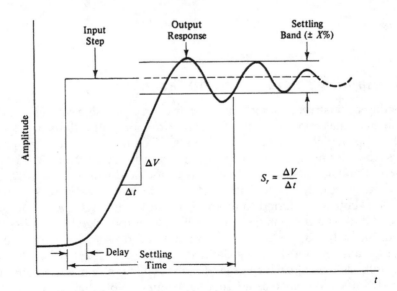

Figure 2.12 Wideband amplifier settling characteristics.

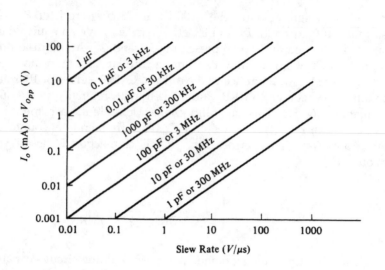

Figure 2.13 Amplifier slew rate curves.

3 MHz typically encountered in video systems specifies an amplifier slew rate of at least 9.45 V/μsec. If the amplifier is also loaded by 1000 pF of capacitance, then it must also be capable of delivering 10 mA of current output at that frequency. These relationships are described by equation (2.22) and its nomograph of Figure 2.13.

$$\mathrm{Sr} = V_{o_{pp}} \bullet \pi \bullet f_{\mathrm{signal}} = \frac{I_o}{C_{sh}} \ \mathrm{V/s} \qquad (2.22)$$

2.5 Amplifier parameter error evaluation

Selecting an instrumentation amplifier involves choosing amplifier input parameters that minimize amplification errors for applications of interest. Table 2.3 provides an error comparison between the five diverse amplifier types, considering application-specific V_{cm} and R_s input values, with evaluation of voltage offsets, interference rejection, and gain nonlinearity. The individual error totals in Table 2.4 provide a performance summary expressed both as a referred-to-input (RTI) amplitude-threshold uncertainty in volts and as a percentage of the full-scale output signal $V_{O_{FS}}$ following amplification by Av_{diff} . Error totals are derived from respective amplifier input parameter contributions defined in equation (2.23), where barred quantities denote mean values and unbarred quantities systematic and random values combined as the root-sum-square. Note that Av_{diff} normally is scaled for the V_{diff} input signal maximum in order to achieve a $V_{O_{FS}}$ of

Table 2.4 Amplifier Error Comparison

	OP-07	OPA-128	AD624	AD215	OPA-646	Comment
R_s	10 KΩ	10 MΩ	1 KΩ	50 Ω	75 Ω	Input group
V_{CM}	±10 V	±10 V	±10 V	±1000 V	±10 V	Input group
V_{OS}	$\overline{10}$ μV	$\overline{140}$ μV	$\overline{25}$ μV	$\overline{400}$ μV	$\overline{1000}$ μV	Offset group
$\dfrac{dV_{OS}}{dT}\bullet dT$	2 μV	50 μV	2.5 μV	20 μV	120 μV	Offset group
$I_{OS}\bullet R_s$	$\overline{3}$ μV	$\overline{0.3}$ μV	$\overline{10}$ μV	$\overline{15}$ μV	$\overline{30}$ μV	Offset group
$6.6\,Vn\sqrt{f_{hi}}$	51 μV	126 μV	26 μV	Negligible	1262 μV	Interference group
$\dfrac{V_{CM}}{CMRR}$	1000 μV	1000 μV	100 μV	10,000 μV	1000 μV	Interference group
$f(Av)\bullet\dfrac{Vo_{FS}}{Av_{diff}}$	$\overline{100}$ μV	$\overline{100}$ μV	$\overline{10}$ μV	$\overline{50}$ μV	$\overline{250}$ μV	Nonlinearity group
$\dfrac{dAv}{dT}\bullet dT\bullet\dfrac{Vo_{FS}}{Av_{diff}}$	500 μV	500 μV	50 μV	150 μV	500 μV	Nonlinearity group
$\varepsilon_{ampl\,RTI}$	$(\overline{113}+1119)$μV	$(\overline{240}+1126)$μV	$(\overline{45}+115)$μV	$(\overline{465}+10{,}003)$μV	$(\overline{1280}+1690)$μV	$\sum\overline{mean}+1\sigma RSS$
$\varepsilon_{ampl\%FS}$	0.123%FS	0.136%FS	0.016%FS	1.046%FS	0.297%FS	$\dfrac{Av_{diff}}{Vo_{FS}}\times\quad\bullet\,100\%$

Note: $V_{diff}=1$ V, $Av_{diff}=1$, $Vo_{FS}=1$ V, $dT=10°C$.

interest at the amplifier output. However, for the normalized examples of Table 2.4, each Av_{diff} is unity, requiring input V_{diff} values that equal the Vo_{FS} value.

$$\varepsilon_{ampl\%FS} = \{\varepsilon_{amplRTI\ volts}\} \times \frac{Av_{diff}}{Vo_{FS}} \bullet 100\% \tag{2.23}$$

$$= \left\{ \overline{V_{OS}} + \overline{I_{OS} \bullet R_S} + \overline{f(A_v) \bullet \frac{Vo_{FS}}{Av_{diff}}} \right.$$

$$+ \left[\left(\frac{dV_{OS}}{dT} \bullet dT \right)^2 + \left(\frac{V_{CM}}{CMRR} \right)^2 + (6.6\ V_n \sqrt{f_{hi}})^2 \right.$$

$$\left. \left. + \left(\frac{dAv}{dT} \bullet dT \bullet \frac{Vo_{FS}}{Av_{diff}} \right)^2 \right]^{1/2} \right\} \times \frac{Av_{diff}}{Vo_{FS}} \bullet 100\%$$

Each amplifier is evaluated at identical Av_{diff}, Vo_{FS}, and temperature, dT, for consistency, but at expected R_s and V_{cm} values relevant to their typical application. All of the amplifiers are capable of accommodating off-ground and electromagnetically coupled V_{cm} input interference with an effectiveness determined by their respective CMRR, where the influence of amplifier CMRR values in attenuating respective V_{cm} values is described. Mean offset voltages, V_{os}, are also untrimmed to reveal these possible differences. The OP-07 is assumed applied to an austere four-resistor subtractor circuit resulting in its 10-K R_s, whereas the OPA-128 low-input-bias-current amplifier interfaces a 10-M R_s current sensor. The AD624 three-amplifier circuit offers the best performance and robustness overall, with gain nonlinearity values a tenth that of the other amplifiers, all of which are normalized to amplifier inputs by the ratio Vo_{FS}/Av_{diff}.

The AD215 isolation amplifier 50-Ω R_s represents either the output of a preceding front-end instrumentation amplifier or low-level emf sensor. The presence of a 1000-volt V_{cm} input essentially accounts for the total error of this amplifier, which will be safely accommodated by the amplifier physical structure. Finally, with a 75-Ω coax R_s, the wideband OPA-646 differs from other amplifiers in providing 10 times the bandwidth at 10 times the internal noise contribution. All amplifier error totals are commensurable owing to like manufacturing technologies. Amplifier V_n rms internal noise voltage is converted to peak-peak with multiplication by 6.6, to account for its crest factor, dimensionally equivocating it to other amplifier input values in each error total.

Bibliography

1. Bailey, D.C., "An Instrumentation Amplifier Is Not an Op Amp," *Electronic Products*, September 18, 1972.
2. Connelly, J.A., *Analog Integrated Circuits*, Wiley Interscience, New York, 1975.

3. Embinder, J., *Application Considerations for Linear Integrated Circuits*, Wiley Interscience, New York, 1970.

4. Fitchen, F.C., *Electronic Integrated Circuits and Systems*, Van Nostrand Reinhold, New York, 1970.

5. Garrett, P.H., *Analog I/O Design: Acquisition — Conversion — Recovery*, Reston Publishing Co., Reston, VA, 1981.

6. Graeme, J.G., *Applications of Operational Amplifiers: Third-Generation Techniques*, McGraw-Hill, New York, 1973.

7. Jaquay, J.W., "Designer's Guide to Instrumentation Amplifiers," *Electronic Design News*, May 5, 1972, p. 40.

8. Kollataj, J.H., "Reject Common-Mode Noise," *Electronic Design*, April 26, 1973, p. 120.

9. Lyerly, T.C., "Instrumentation Amplifier Conditions Computer Inputs," *Electronics*, November 6, 1972, p. 115.

10. Netzer, Y., "The Design of Low-Noise Amplifiers," *Proceedings IEEE*, June 1981.

11. Pettit, J.M. and McWhorter, M.M., *Electronic Amplifier Circuits*, McGraw-Hill, New York, 1961.

12. Rutkowski, G.B., *Handbook of Integrated-Circuit Operational Amplifiers*, Prentice Hall, Englewood Cliffs, NJ, 1975.

13. Tobey, G., Graeme, J., and Huelsman, L., *Operational Amplifiers: Design and Applications*, McGraw-Hill, New York, 1971.

14. Young, R.L., "Lift IC Op-Amp Performance," *Electronic Design*, February 15, 1973, p. 66.

chapter three

Instrumentation filters with nominal error

3.0 Introduction

Although the requirement for electric wave filters extends over an interval of a century to Marconi's radio experiments, the identification of stable and ideally terminated filter networks has occurred only during the past 25 years. Filtering at the lower instrumentation frequencies has always been a problem with passive filters because the required L and C values are larger and inductor losses appreciable. The bandlimiting of measurement signals in instrumentation applications imposes the additional concern of filter error additive to these measurement signals when accurate signal processing is required.

Consequently, this chapter provides a development of lowpass and bandpass filter characterizations appropriate for measurement signals and develops filter error analyses for the more frequently required lowpass realizations. The excellent stability of active filter networks in the dc-to-100-kHz instrumentation frequency range makes these circuits especially useful. Filter error analysis is accordingly developed to optimize the implementation of filters for input signal conditioning, aliasing prevention, and output interpolation purposes associated with data conversion systems.

3.1 Bandlimiting instrumentation filters

Lowpass filters are frequently required to bandlimit measurement signals in instrumentation applications to achieve a frequency-selective function of interest. The application of an arbitrary signal set to a lowpass filter can result in a significant attenuation of higher frequency components, thereby defining a stopband with a boundary influenced by the choice of filter cutoff frequency, with the unattenuated frequency components defining the filter passband. For instrumentation purposes, approximating the ideal lowpass filter amplitude $A(f)$ and phase $B(f)$ responses, described by Figure 3.1, is beneficial in order to achieve signal bandlimiting without alteration or the addition of errors to a passband signal of interest. In fact, preserving the

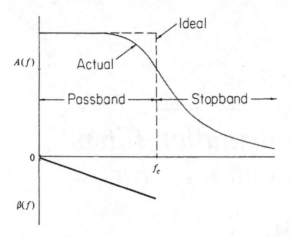

Figure 3.1 Ideal lowpass filter.

accuracy of measurement signals is of sufficient importance that consideration of filter characterizations that correspond to well-behaved functions such as Butterworth and Bessel polynomials are especially useful. However, an ideal filter is physically unrealizable because practical filters are represented by ratios of polynomials that cannot possess the discontinuities required for sharply defined filter boundaries.

Figure 3.2 describes the Butterworth lowpass amplitude response, $A(f)$, and Figure 3.3 its phase response, $B(f)$, where n denotes the filter order or

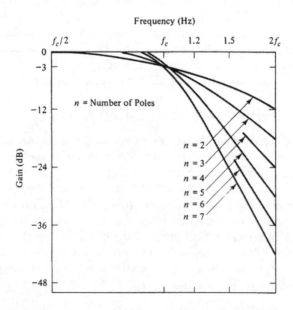

Figure 3.2 Butterworth lowpass amplitude.

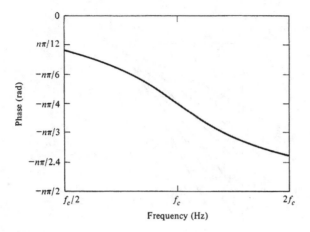

Figure 3.3 Butterworth lowpass phase.

number of poles. Butterworth filters are characterized by a maximally flat amplitude response in the vicinity of dc, which extends toward its –3-dB cutoff frequency, f_c, as n increases. This characteristic is defined by equations (3.1) and (3.2) and Table 3.1.

$$A(f) = \frac{b_o}{\sqrt{B(s)B(-s)}} = \frac{1}{\sqrt{1 + \left(\dfrac{f}{f_c}\right)^{2n}}} \qquad (3.1)$$

$$B(s) = \left(j\frac{f}{f_c}\right)^n + b_{n-1}\left(j\frac{f}{f_c}\right)^{n-1} + \ldots + b_o \qquad (3.2)$$

Butterworth attenuation is rapid beyond f_c as filter order increases with a slightly nonlinear phase response that provides a good approximation to an ideal lowpass filter. An analysis of the error attributable to this approximation is derived in Section 3.3. Figure 3.4 presents the Butterworth highpass response.

Table 3.1 Butterworth Polynomial Coefficients

Poles, n	b_0	b_1	b_2	b_3	b_4	b_5
1	1.0					
2	1.0	1.414				
3	1.0	2.0	2.0			
4	1.0	2.613	3.414	2.613		
5	1.0	3.236	5.236	5.236	3.236	
6	1.0	3.864	7.464	9.141	7.464	3.864

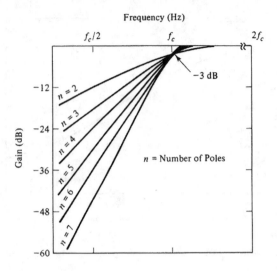

Figure 3.4 Butterworth highpass amplitude.

Bessel filters are allpole filters, like their Butterworth counterparts, with an amplitude response described by equations (3.3) and (3.4) and Table 3.2.

$$A(f) = \frac{b_o}{\sqrt{B(s)B(-s)}}$$ (3.3)

$$B(s) = \left(j\frac{f}{f_c}\right)^n + b_{n-1}\left(j\frac{f}{f_c}\right)^{n-1} + \ldots + b_o$$ (3.4)

Bessel lowpass filters are characterized by a more linear phase delay extending to their cutoff frequency, f_c, and beyond as a function of filter order, n, shown in Figure 3.5. However, this linear phase property applies only to lowpass filters. Unlike the flat passband of Butterworth lowpass filters, the Bessel passband has no value that does not exhibit amplitude attenuation with a Gaussian amplitude response, described by Figure 3.6. It is also useful to compare the overshoot of Bessel and Butterworth filters in Table 3.3, which reveals the Bessel to be much better behaved for bandlimiting pulse-type instrumentation signals and where phase linearity is essential.

Table 3.2 Bessel Polynomial Coefficients

Poles, n	b_0	b_1	b_2	b_3	b_4	b_5
1	1					
2	3	3				
3	15	15	6			
4	105	105	45	10		
5	945	945	420	105	15	
6	10,395	10,395	4725	1260	210	21

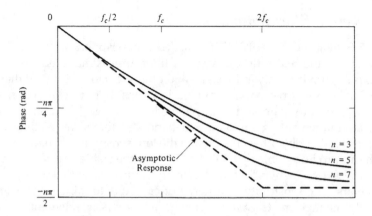

Figure 3.5 Bessel lowpass phase.

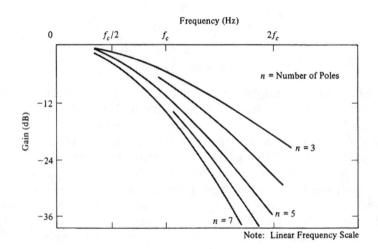

Figure 3.6 Bessel lowpass amplitude.

Table 3.3 Filter Overshoot Pulse Response

n	Bessel (%FS)	Butterworth (%FS)
1	0	0
2	0.4	4
3	0.7	8
4	0.8	11

3.2   Active filter networks

In 1955 Sallen and Key of MIT published a description of 18 active filter networks for the realization of various filter approximations. However, a rigorous sensitivity analysis by Geffe and others disclosed by 1970 that only four of the original networks exhibited low sensitivity to component drift. Of these, the unity gain and multiple-feedback networks are of particular value for implementing lowpass and bandpass filters, respectively, to Q values of 10. Work by others resulted in the low-sensitivity biquad resonator, which can provide stable Q values to 200, and the stable gyrator bandreject filter. These four networks are shown in Figure 3.7 with key sensitivity parameters. The sensitivity of a network can be determined, for example, when the change in its Q for a change in its passive-element values is evaluated. Equation (3.5) describes the change in the Q of a network by multiplying the thermal coefficient of the component of interest by its sensitivity coefficient. Normally, 50-to-100-ppm/°C components yield good performance.

$$S_Z^Q = \pm 1 \text{ passive network}$$
$$= (\pm 1)(50 \text{ ppm}/°C)(100\%) \qquad (3.5)$$
$$= \pm 0.005\% \ Q/°C$$

Unity gain networks offer excellent performance for lowpass and highpass realizations and may be cascaded for higher order filters. This is perhaps the most widely applied active filter circuit. Note that its sensitivity coefficients are less than unity for its passive components — the sensitivity of conventional passive networks — and that its resistor temperature coefficients are zero. However, it is sensitive to filter gain, indicating that designs that also obtain greater than unity gain with this filter network are suboptimum. The advantage of the multiple-feedback network is that a bandpass filter can be formed with a single operational amplifier, although the biquad network must be used for high-Q bandpass filters. However, the stability of the biquad at higher Q values depends on the availability of adequate amplifier loop gain at the filter center frequency. Both bandpass networks can be stagger tuned for a maximally flat passband response when required. The principle of operation of the gyrator is that a negative conductance, G gyrates a capacitive current to an effective inductive current. Frequency stability is very good, and a bandreject filter notch depth to about −40 dB is generally available. It should be appreciated that the principal capability of the active filter network is to synthesize a complex–conjugate pole pair. This achievement, as described below, permits the realization of any mathematically definable lowpass approximation and is focused on application of unity gain networks in this chapter.

Kirchoff's current law provides that the sum of the currents into any node is zero. A nodal analysis of the unity gain lowpass network yields equations (3.6) through (3.9). It includes the assumption that current in C_2 is equal to current in R_2; the realization of this requires the use of a

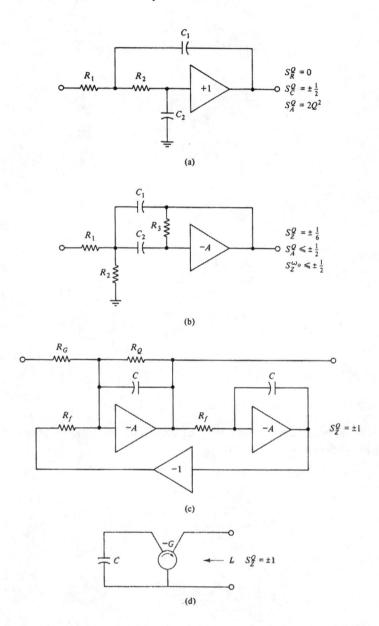

(a)

(b)

(c)

(d)

Figure 3.7 Recommended active filter networks: (a) unity gain, (b) multiple feedback, (c) biquad, and (d) gyrator.

low-input-bias-current operational amplifier for accurate performance. The transfer function is obtained upon substituting for V_x in equation (3.6) its independent expression obtained from equation (3.7). Filter pole positions are defined by equation (3.9). Figure 3.8 shows these nodal equations and the complex-plane pole positions mathematically described by equation (3.9).

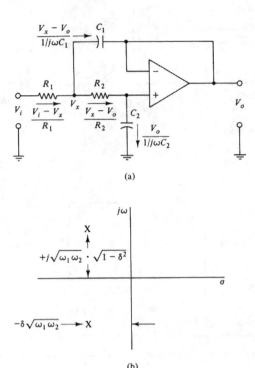

(a)

(b)

Figure 3.8 Unity gain network nodal analysis.

This second-order network has two denominator roots (two poles) and is sometimes referred to as a resonator.

$$\frac{V_i - V_x}{R_1} = \frac{V_x - V_o}{1/j\omega C_1} + \frac{V_x - V_o}{R_2} \tag{3.6}$$

$$\frac{V_x - V_o}{R_2} = \frac{V_o}{1/j\omega C_2} \tag{3.7}$$

Rearranging,

$$V_x = V_o \bullet \frac{R_2 + 1/j\omega C_2}{1/j\omega C_2}$$

$$\frac{V_0}{V_i} = \frac{1}{\omega^2 R_1 R_2 C_1 C_2 + \omega C_2 R_1 + R_2 + 1} \tag{3.8}$$

$$\omega_1 = \frac{1}{R_1 C_1} \quad \text{and} \quad \omega_2 = \frac{1}{R_2 C_2}$$

$$\delta = \frac{C_2}{2}(R_1 + R_2)$$

$$S_{1,2} = \delta\sqrt{\omega_1\omega_2} \pm j\sqrt{\omega_1\omega_2} - \sqrt{1-\delta^2} \tag{3.9}$$

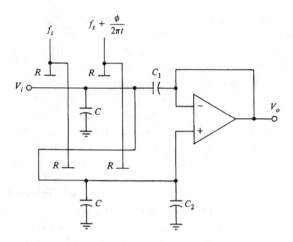

Figure 3.9 Switched-capacitor unity gain network.

A recent technique using metal oxide semiconductor (MOS) technology has made possible the realization of multipole unity gain network active filters in total integrated-circuit form without the requirement for external components. Small-value MOS capacitors are utilized with MOS switches in a switched-capacitor circuit for simulating large-value resistors under control of a multiphase clock. With reference to Figure 3.9, the rate, f_s, at which the capacitor is toggled determines its charging to V and discharging to V'. Consequently, the average current flow, I, from V to V' defines an equivalent resistor, R, that would provide the same average current shown by the identity of equation (3.10).

$$R = \frac{V - V'}{I} = 1/Cf_c \qquad (3.10)$$

The switching rate, f_s, is normally much higher than the signal frequencies of interest so that the time sampling of the signal can be ignored in a simplified analysis. Filter accuracy is primarily determined by the stability of the frequency, f_s, and the accuracy of implementation of the monolithic MOS capacitor ratios.

The most important parameter in the selection of operational amplifiers for active filter service is open-loop gain. The ratio of open-loop to closed-loop gain, or loop gain, must be 100 or greater for stable and well-behaved performance at the highest signal frequencies present. This is critical in the application of bandpass filters to ensure a realization that accurately follows the design calculations. Amplifier input and output impedances are normally sufficiently close to the ideal infinite input and zero output values to be inconsequential for impedances in active filter networks. Metal film resistors having a temperature coefficient of 50 ppm/°C are recommended for active filter design.

Selection of capacitor type is the most difficult decision because of many interacting factors. For most applications, polystyrene capacitors are recommended because of their reliable −120 ppm/°C negative temperature coefficient and 0.05% capacitance retrace deviation with temperature cycling. Where capacitance values above 0.1 μF are required, however, polycarbonate capacitors are available in values to 1 μF, with a ±50 ppm/°C temperature coefficient and 0.25% retrace. Mica capacitors are the most stable devices, with ±50 ppm/°C temperature coefficient and 0.1% retrace, but practical capacitance availability is typically only 100 to 5000 pF. Mylar capacitors are available in values to 10 μF, with 0.3% retrace, but their temperature coefficient averages 400 ppm/°C.

The choice of resistor and capacitor tolerance determines the accuracy of the filter implementation such as its cutoff frequency and passband flatness. Cost considerations normally dictate the choice of 1% tolerance resistors and 2 to 5% tolerance capacitors. However, usual practice is to pair larger and smaller capacitor values to achieve required filter network values to within 1%, which results in filter parameters accurate to 1 or 2% with low temperature coefficient and retrace components. Filter response is typically displaced inversely to passive-component tolerance, such as lowering of cutoff frequency for component values on the high side of their tolerance band. For more critical realizations, such as high-Q bandpass filters, some provision for adjustment provides the flexibility needed for an accurate implementation.

Table 3.4 provides the capacitor values in farads for unity gain networks tabulated according to the number of filter poles. Higher-order filters are

Table 3.4 Unity Gain Network Capacitor Values in Farads

Poles	Butterworth			Bessel		
	C_1	C_2	C_3	C_1	C_2	C_3
2	1.414	0.707		0.907	0.680	
3	3.546	1.392	0.202	1.423	0.988	0.254
4	1.082	0.924		0.735	0.675	
	2.613	0.383		1.012	0.390	
5	1.753	1.354	0.421	1.009	0.871	0.309
	3.235	0.309		1.041	0.310	
6	1.035	0.966		0.635	0.610	
	1.414	0.707		0.723	0.484	
	3.863	0.259		1.073	0.256	
7	1.531	1.336	0.488	0.853	0.779	0.303
	1.604	0.624		0.725	0.415	
	4.493	0.223		1.098	0.216	
8	1.091	0.981		0.567	0.554	
	1.202	0.831		0.609	0.486	
	1.800	0.556		0.726	0.359	
	5.125	0.195		1.116	0.186	

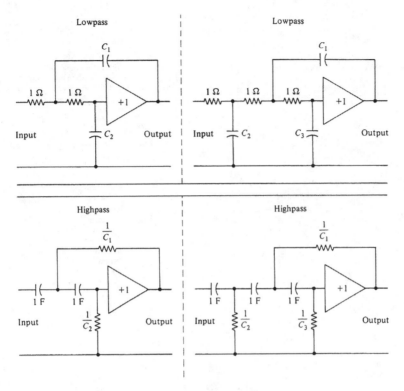

Figure 3.10 Two- and three-pole unity gain networks.

formed by a cascade of the second- and third-order networks shown in Figure 3.10, each of which is different. For example, a sixth-order filter will have six different capacitor values and not consist of a cascade of identical two-pole or three-pole networks. Figures 3.11 and 3.12 illustrate the design procedure with 1-kHz-cutoff two-pole Butterworth lowpass and highpass filters, respectively, including the frequency and impedance scaling steps. The three-pole filter design procedure is identical with observation of the appropriate network capacitor locations but should be driven from a low driving-point impedance such as an operational amplifier. A design guide for unity gain active filters is summarized in the following steps:

1. Select an appropriate filter approximation and number of poles required to provide the necessary response from the curves of Figures 3.2 through 3.6.
2. Choose the filter network appropriate for the required realization from Figure 3.10 and perform the necessary component frequency and impedance scaling.
3. Implement the filter components by selecting 1% standard-value resistors and then pairing a larger and smaller capacitor to realize each capacitor value to within 1%.

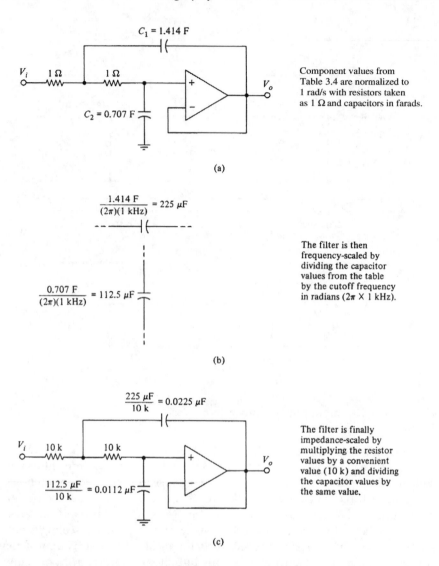

Component values from Table 3.4 are normalized to 1 rad/s with resistors taken as 1 Ω and capacitors in farads.

The filter is then frequency-scaled by dividing the capacitor values from the table by the cutoff frequency in radians ($2\pi \times 1$ kHz).

The filter is finally impedance-scaled by multiplying the resistor values by a convenient value (10 k) and dividing the capacitor values by the same value.

Figure 3.11 Butterworth unity gain lowpass filter example.

3.3 Filter error analysis

Requirements for signal bandlimiting in data acquisition and conversion systems include signal quality upgrading by signal conditioning circuits, aliasing prevention associated with sampled-data operations, and intersample error smoothing in output signal reconstruction. The accuracy, stability, and efficiency of lowpass active filter networks satisfy most of these requirements, with the realization of filter characteristics appropriate for specific applications. However, when a filter is superimposed on a signal of interest,

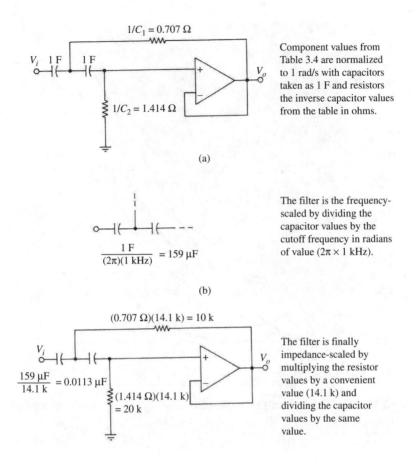

Figure 3.12 Butterworth unity-gain highpass filter example.

filter gain and phase deviations from the ideal result in a signal amplitude error that constitutes component error. It is therefore useful to evaluate filter parameters in order to select filter functions appropriate for signals of interest. It will be shown that applying this approach results in a minimum filter error added to the total system error budget. Because dc, sinusoidal, and harmonic signals are encountered in practice, analysis is performed for these signal types to identify optimum filter parameters for achieving minimum error.

$$\overline{\varepsilon_{\%FS}} = \frac{0.1}{BW/fc} \sum_{o}^{BW/fc} (1.0 - A(f)) \bullet 100\% \qquad \text{dc \& sinusoidal signals} \quad (3.11)$$

Both dc and sinusoidal signals exhibit a single spectral term. Filter gain error is thus the primary source of error because single line spectra are

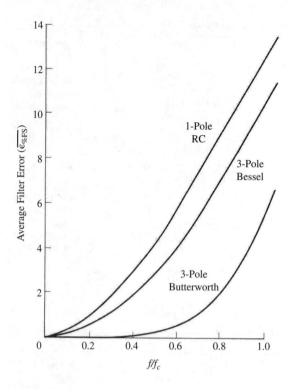

Figure 3.13 Plot of filter errors for dc and sinusoidal signals as a function of passband fraction.

unaffected by filter phase nonlinearities. Figure 3.13 describes the passband gain deviation, with reference to 0 Hz and expressed as average percent error of full scale, for three lowpass filters. The filter error attributable to gain deviation $(1.0 - A(f))$ is shown to be minimum for the Butterworth characteristic, which is an expected result considering the passband flatness provided by Butterworth filters. Of significance is that small filter component errors can be achieved by restricting signal spectral occupancy to a fraction of the filter cutoff frequency.

Table 3.5 presents a tabulation of the example filters evaluated with dc and sinusoidal signals, defining mean amplitude errors for signal bandwidth occupancy to specified filter passband fractions of the cutoff frequency, f_c. Equation (3.11) provides an approximate fit of filter parameters for RC, Bessel, and Butterworth filter characteristics. Most applications are better served by the three-pole Butterworth filter, which offers a component error of 0.1 %FS for signal passband occupancy to 40% of the filter cutoff, plus good stopband attenuation. Although it may appear inefficient not to utilize a filter passband up to its cutoff frequency, the total bandwidth sacrificed is usually small. Higher-order filters may also be evaluated when greater

Table 3.5 Filter Amplitude Errors for DC and Sinusoidal Signals

Signal Bandwidth Passband Fractional Occupancy	Amplitude Response, $A(f)$			Average Filter Error, $\overline{\varepsilon_{\%FS}}$		
$\dfrac{BW}{f_c}$	1-pole RC	3-pole Bessel	3-pole Butterworth	1-pole RC	3-pole Bessel	3-pole Butterworth
0.05	0.999	0.999	1.000	0.1	0.1	0
0.1	0.997	0.998	1.000	0.3	0.2	0
0.2	0.985	0.988	1.000	0.9	0.7	0
0.3	0.958	0.972	1.000	1.9	1.4	0
0.4	0.928	0.951	0.998	3.3	2.3	0.1
0.5	0.894	0.924	0.992	4.7	3.3	0.2
0.6	0.857	0.891	0.977	6.3	4.6	0.7
0.7	0.819	0.852	0.946	8.0	6.0	1.4
0.8	0.781	0.808	0.890	9.7	7.7	2.6
0.9	0.743	0.760	0.808	11.5	9.5	4.4
1.0	0.707	0.707	0.707	13.3	11.1	6.9

stopband attenuation is of interest, with substitution of their amplitude response, $A(f)$, in equation (3.11).

The consequence of nonlinear phase delay with harmonic signals is described by Figure 3.14. The application of a harmonic signal just within the passband of a six-pole Butterworth filter provides the distorted output waveform shown. The variation in time delay between signal components at their specific frequencies results in a signal time displacement and the amplitude alteration described. This time variation is apparent from evaluation of equation (3.12), where linear phase provides a constant time delay. An error signal, $\varepsilon(t)$, is derived as the difference between the output, $y(t)$, of

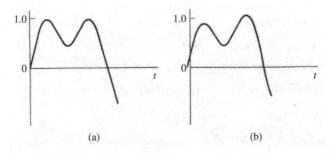

Figure 3.14 Filtered complex waveform phase nonlinearity: (a) sum of fundamental and third harmonic in 2:1 ratio; and (b) sum of fundamental and third harmonic following six-pole lowpass Butterworth filter with signal spectral occupancy to filter cutoff.

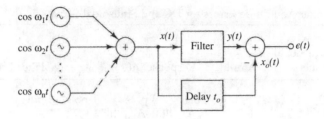

Figure 3.15 Filter harmonic signal error evaluation.

a filter of interest and a delayed input signal, $x_o(t)$, expressed by equations (3.13) through (3.15) and described in Figure 3.15. A volts-squared output error is then obtained from the Fourier transform of this error signal and the application of trigonometric identities; the error is expressed in terms of mean squared error (MSE) by equation (3.16), with A_n and ϕ_n, respectively, the filter magnitude and phase responses at n frequencies.

Computer simulation of first- through eighth-order Butterworth and Bessel lowpass filters were obtained with the structure displayed in Figure 3.15. The signal delay, t_o, was varied in a search for the minimum true MSE by applying the Newton-Raphson method to the derivative of the MSE expression. This exercise was repeated for each filter with various passband spectral occupancies ranging from 10 to 100% of the cutoff frequency and $N = 10$ sinusoids per octave represented as the simulated input signal, $x(t)$. MSE is calculated by the substitution of each t_o value in equation (3.16) and expressed as average filter component error, $\varepsilon_{\%FS}$, by equation (3.17) over the filter passband fraction specified for signal occupancy.

$$\text{Delay variation} = \frac{\phi_a}{2\pi f_a} - \frac{\phi_b}{2\pi f_b} \ \sec \tag{3.12}$$

$$y(t) = \sum_{n=1}^{N} A_n \cos(\omega_n t - \phi_n) \tag{3.13}$$

$$x_o(t) = \sum_{n=1}^{N} \cos(\omega_n t - \omega_n t_o) \tag{3.14}$$

$$\varepsilon(t) = y(t) - x_o(t) = \sum_{n=1}^{N} [A_n \cos(\omega_n t - \phi_n) - \cos(\omega_n t - \omega_n t_o)] \tag{3.15}$$

$$\text{MSE} = \frac{1}{2} \sum_{n=1}^{N} \left[(A_n \cos\phi_n - \cos\omega_n t_o)^2 + (A_n \sin\phi_n - \sin\omega_n t_o)^2 \right] \tag{3.16}$$

$$\varepsilon_{\%FS} = \frac{\sqrt{\text{MSE}}}{x(t)} \bullet 100\% \ \text{harmonic signals} \tag{3.17}$$

Table 3.6 Filter Amplitude Errors for Harmonic Signals

Filter Order (Poles)			Average Filter Error, $\varepsilon_{\%FS}$		
RC	Butterworth	Bessel	$f_c = 10$ BW[a]	$f_c = 3$ BW[a]	$f_c = $ BW[a]
1			1.201%		
	2			1.093%	6.834%
		2		0.688	6.179
	3			0.115	5.287
		3		0.677	6.045
	4			0.119	5.947
		4		0.698	6.075
	5			0.134	6.897
		5		0.714	6.118
	6			0.153	7.900
		6		0.725	6.151
	7			0.172	8.943
		7		0.997	6.378
	8			0.195	9.996
		8		1.023	6.299

[a] BW = Signal bandwidth.

Table 3.6 presents these results with an efficient filter-cutoff-to-signal-bandwidth ratio, $f_c/$BW, of 3, considering filter passband signal occupancy vs. minimized filter error. Signal spectral occupancy up to the filter cutoff frequency is also simulated for error reference purposes. The application of higher-order filters is primarily determined by the need for increased stopband attenuation compared with the additional complexity and component precision required for their realization.

Lowpass bandlimiting filters are frequently required by signal conditioning channels, as illustrated in the following chapters, and especially for presampling antialiasing purposes plus output signal interpolation in sampled-data systems. Of interest is whether the intelligence represented by a signal is encoded in its amplitude values, phase relationships, or both. Filter mean nonlinearity errors presented in Tables 3.5 and 3.6 describe amplitude deviations of filtered signals resulting from nonideal filter magnitude and phase characteristics. It is clear from these tabulations that Butterworth filters contribute nominal error-to-signal amplitudes when their passband cutoff frequency is derated to multiples of a signal bandwidth value. Also notable is that measurands and encoded data are so commonly represented by signal amplitude values in instrumentation systems that Butterworth filters predominate.

When signal phase accuracy is essential for phase-coherent applications, ranging from communications to audio systems including matrixed home theater signals, then Bessel lowpass filters are advantageous. For example, if only signal phase is of interest, an examination of Figure 3.5 and Tables 3.5 and 3.6 reveals that derating a three-pole Bessel filter passband cutoff frequency to three times the signal bandwidth achieves very linear phase, but signal amplitude error approaches 1%FS. However, error down to 0.1 – 0.2%FS in both amplitude and phase are provided for any signal type when this lowpass filter is derated on the order of ten times signal bandwidth. At that operating point Bessel filters behave as a pure delay line to the signal.

Bibliography

1. Allen, P.E. and Huelsman, L.P., *Theory and Design of Active Filters*, John Wiley, New York, 1975.
2. Brockman, J.P., "Interpolation Error in Sampled Data Systems," Electrical Engineering Department, University of Cincinnati, May 1985.
3. Craig, J.W., *Design of Lossy Filters*, MIT Press, Cambridge, MA, 1970.
4. Daniels, R.W., *Approximation Methods for Electronic Filter Design*, McGraw-Hill, New York, 1974.
5. Geffe, P.R., "Toward High Stability in Active Filters," *IEEE Spectrum*, 7, May 1970.
6. Johnson, D.E., *Introduction to Filter Theory*, Prentice Hall, Englewood Cliffs, NJ, 1976.
7. Johnson, D.E. and Hilburn, J.L., *Rapid Practical Designs of Active Filters*, John Wiley, New York, 1975.
8. Laube, S., "Comparative Analysis of Total Average Filter Component Error," Senior Design Project, Electrical Engineering Technology, University of Cincinnati, 1983.
9. Mitra, C., *Analysis and Synthesis of Linear Active Networks*, John Wiley, New York, 1969.
10. Rhodes, J.D., *Theory of Electrical Filters*, John Wiley, New York, 1976.
11. Sallen, R.P. and Key, E.L., "A Practical Method of Designing RC Active Filters," *IRE Transactions on Circuit Theory*, CT-2, March 1955.
12. Thomas, L.C., "The Biquad, Part 1 — Some Practical Design Considerations," *IEEE Circuit Theory Transactions*, CT-18, May 1971.
13. Zeines, B., *Introduction to Network Analysis*, Prentice Hall, Englewood Cliffs, NJ, 1967.

chapter four

Signal acquisition, conditioning, and processing

4.0 Introduction

Economic considerations are imposing increased accountability on designers of analog I/O systems to provide performance at the required accuracy for computer-integrated measurement and control instrumentation without the costs of overdesign. Within that context this chapter provides the development of signal acquisition and conditioning systems and derives a unified method for representing and upgrading the quality of instrumentation signals between sensors and data-conversion systems. Low-level signal conditioning is comprehensively developed for both coherent and random interference conditions employing sensor–amplifier–filter structures for signal quality improvement presented in terms of detailed device and system error budgets. Examples for dc, sinusoidal, and harmonic signals are provided, including grounding, shielding, and noise circuit considerations. A final section explores the additional signal quality improvement available by averaging redundant signal conditioning channels plus extension to analog signal processing. A distinction is made between signal conditioning, which is primarily concerned with operations for improving signal quality, and signal processing operations that assume signal quality at the level of interest.

4.1 Low-level signal acquisition

Designers of high-performance instrumentation systems must define criteria for determining preferred options from available alternatives. Figure 4.1 illustrates a cause-and-effect outline of methods developed in this chapter with applications that aid the realization of effective signal conditioning designs. In this fishbone chart, grouped system and device options are outlined for contributing to the goal of minimum total instrumentation error.

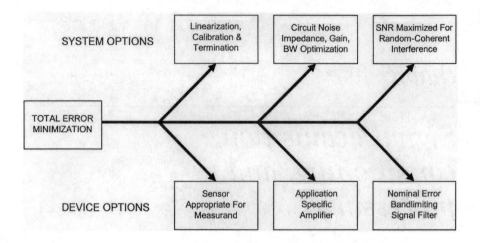

Figure 4.1 Signal conditioning design influences.

Sensor choices appropriate for measurands of interest, including linearization and calibration, were introduced in Chapter 1. Application-specific amplifier and filter choices for signal conditioning were defined, respectively, in Chapters 2 and 3. In this section, input circuit noise, impedance, and grounding effects are described for signal conditioning optimization. The following section derives models that combine device and system elements for the evaluation and improvement of signal quality, expressed as total error, arising from the influence of random and coherent interference. The remaining sections provide detailed examples of these signal conditioning design methods.

External interference entering low-level instrumentation circuits frequently is substantial and techniques for its attenuation are essential. Noise coupled to signal cables and power buses has as its cause electric and magnetic field sources. For example, signal cables will couple 1 mV of interference per kilowatt of 60-Hz load for each lineal foot of cable run of 1-ft spacing from adjacent power cables. Most interference results from near-field sources, primarily electric field, whereby an effective attenuation mechanism is reflection by nonmagnetic materials such as copper or aluminum shielding. Both foil and braided shielded twinax signal cables offer attenuation on the order of –90 voltage dB to 60-Hz interference, which degrades by approximately +20 dB per decade of increasing frequency.

For magnetic fields absorption is the effective attenuation mechanism, requiring steel or mu-metal shielding. Magnetic fields are more difficult to shield than electric fields, where shielding effectiveness for a specific thickness diminishes with decreasing frequency. For example, at 60 Hz steel provides interference attenuation on the order of –30 voltage dB per 100 mil of thickness. Applications requiring magnetic shielding are usually implemented by the installation of signal cables in steel conduit of the necessary wall thickness. Additional magnetic-field attenuation is furnished by periodic transposition of twisted-pair signal cable, provided no signal returns are on

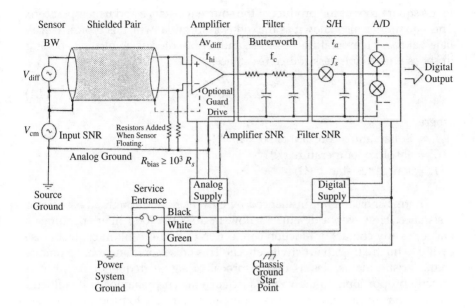

Figure 4.2 Signal acquisition system structure.

the shield, where low-capacitance cabling is preferable. Mutual coupling between computer data-acquisition system elements — for example from finite ground impedances shared among different circuits — also can be significant, with noise amplitudes equivalent to 50 mV at signal inputs. Such coupling is minimized by separating analog and digital circuit grounds into separate returns to a common low-impedance chassis star-point termination, as illustrated in Figure 4.2.

The goal of shield ground placement in all cases is to provide a barrier between signal cables and external interference from sensors to their amplifier inputs. Signal cable shields also are grounded at a single point, below 1 MHz signal bandwidths, and ideally at the source of greatest interference where provision of the lowest impedance ground is more beneficial. One instance in which a shield is not grounded is when driven by an amplifier guard. Guarding neutralizes cable-to-shield capacitance imbalance by driving the shield with common-mode interference appearing on the signal leads, and also is known as active shielding.

The components of total input noise may be divided into external contributions associated with the sensor circuit and internal amplifier noise sources referred to its input. We shall consider the combination of these noise components in the context of bandlimited sensor–amplifier signal acquisition circuits. Phenomena associated with the measurement of a quantity frequently involve energy–matter interactions that result in additive noise. Thermal noise, V_t, is present in all elements containing resistance above absolute zero temperature. Equation (4.1) defines thermal noise voltage proportional to the square root of the product of the source resistance and its temperature. This equation is also known as the Johnson formula, which is typically evaluated at room temperature, or 293 K, and represented as a voltage generator in series with a noiseless source resistance.

$$V_t = \sqrt{4kTR_s} \ \text{Vrms}/\sqrt{\text{Hz}} \qquad (4.1)$$

where
  k = Boltzmann's constant (1.38×10^{-23} J/K)
  T = absolute temperature (K)
  R_s = source resistance (Ω)

Thermal noise is not influenced by current flow through an associated resistance. However, a dc current flow in a sensor loop may encounter a barrier at any contact or junction connection that can result in contact noise owing to fluctuating conductivity effects. This noise component has a unique characteristic that varies as the reciprocal of signal frequency, $1/f$, but is directly proportional to the value of dc current. The behavior of this fluctuation with respect to a sensor-loop source resistance is to produce a contact noise voltage the magnitude of which may be estimated at a signal frequency of interest by the empirical relationship of equation (4.2). An important conclusion is that dc current flow should be minimized in the excitation of sensor circuits, especially for low signal frequencies.

$$V_c = (0.57 \times 10^{-9}) \ R_s \ \sqrt{\frac{I_{dc}}{f}} \ \text{Vrms}/\sqrt{\text{Hz}} \qquad (4.2)$$

where
  I_{dc} = average dc current (A)
  f = signal frequency (Hz)
  R_s = source resistance (Ω)

Instrumentation amplifier manufacturers use the method of equivalent noise–voltage and noise–current sources applied to one input to represent internal noise sources referred to amplifier input, as illustrated in Figure 4.3. The short-circuit rms input noise voltage, V_n, is the random disturbance that would appear at the input of a noiseless amplifier, where its increase below 100 Hz is due to internal amplifier, $1/f$, contact noise sources. The open-circuit rms input noise current, I_n, similarly arises from internal amplifier noise sources and usually may be disregarded in sensor–amplifier circuits because its small magnitude typically results in a negligible input disturbance, except when large source resistances are present. Because all of these input noise contributions are essentially from uncorrelated sources, they are combined as the root sum square (RSS) by equation (4.3). Wide bandwidths and large source resistances, therefore, should be avoided in sensor–amplifier signal acquisition circuits in the interest of noise minimization. Further, additional noise sources

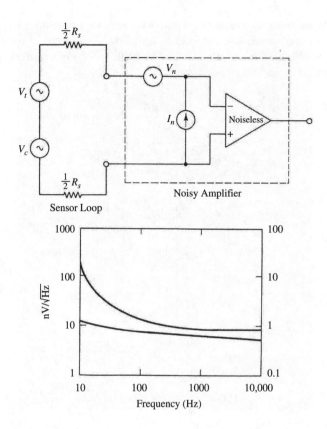

Figure 4.3 Sensor–amplifier noise sources.

encountered in an instrumentation channel following the input gain stage are of diminished consequence because of noise amplification provided by the input stage.

$$V_{N_{pp}} = \left[6.6 \left(V_t^2 + V_c^2 + V_n^2 \right) (f_{hi}) \right]^{1/2} \tag{4.3}$$

4.2   Signal quality in random and coherent interference

The acquisition of a low-level analog signal that represents some measurand in the presence of appreciable interference is a frequent requirement. Of concern is achieving a signal amplitude measurement A or phase angle ϕ at the accuracy of interest through upgrading the quality of the signal by means of appropriate signal conditioning circuits. Closed-form expressions are available for determining the error of a signal corrupted by random Gaussian noise or coherent sinusoidal interference. These are expressed in terms of signal-to-noise ratios (SNRs) by equations (4.4) through (4.9). SNR is a dimensionless ratio of watts of signal to watts of noise and frequently is expressed as rms signal-to-interference amplitude squared. These equations are exact for sinusoidal signals, which are typical for excitation encountered with instrumentation sources.

$$P(\Delta A;\ A) = \mathrm{erf}\left(\frac{1}{2}\frac{\Delta A}{A}\sqrt{\mathrm{SNR}}\right) \text{probability} \qquad (4.4)$$

$$0.68 = \mathrm{erf}\left(\frac{1}{2}\frac{\varepsilon_{\%FS}}{100\%}\sqrt{\mathrm{SNR}}\right)$$

$$\varepsilon_{\text{random amplitude}} = \frac{\sqrt{2}\ 100\%}{\sqrt{\mathrm{SNR}}} \quad \text{of full scale } (1\sigma) \qquad (4.5)$$

$$P(\Delta\varphi;\ \varphi) = \mathrm{erf}\left(\frac{1}{2}\frac{\Delta\varphi}{\varphi}\sqrt{\mathrm{SNR}}\right) \text{probability} \qquad (4.6)$$

$$0.68 = \mathrm{erf}\left(\frac{1}{2}\frac{\varepsilon_{\%FS}}{57.3^{\circ}/\mathrm{rad}}\sqrt{\mathrm{SNR}}\right)$$

$$\varepsilon_{\text{random phase}} = \frac{\sqrt{2}\ 100\%}{\sqrt{\mathrm{SNR}}} \quad \text{degrees } (1\sigma) \qquad (4.7)$$

$$\varepsilon_{\text{coh amplitude}} = \frac{\Delta A}{A}\bullet 100\% = \sqrt{\frac{V_{coh}^{2}}{V_{FS}^{2}}}\bullet 100\% = \frac{100\%}{\sqrt{\mathrm{SNR}}} \quad \text{of full scale} \qquad (4.8)$$

$$\varepsilon_{\text{coh phase}} = \frac{100\%}{2\sqrt{\mathrm{SNR}}} \quad \text{degrees} \qquad (4.9)$$

The probability that a signal corrupted by random Gaussian noise is within a specified Δ region centered on its true amplitude, A, or phase, ϕ, values is defined by equations (4.4) and (4.6). Table 4.1 presents a tabulation from substitution into these equations for amplitude and phase errors at a 68% (1σ) confidence in their measurement for specific SNR values. One sigma is an acceptable confidence level for many applications. For 95% (2σ) confidence, the error values are doubled for the same SNR. These amplitude and phase errors are closely approximated by the simplifications of equations (4.5) and (4.7) and are more readily evaluated than by equations (4.4) and (4.6). For coherent interference, equations (4.8) and (4.9) approximate amplitude and phase errors, where ΔA is directly proportional to V_{coh} as the true value of A is to V_{FS}. Errors due to coherent interference are seen to be less than those due to random interference by the $\sqrt{2}$ for identical SNR values. Further, the accuracy of all of these error expressions requires minimum SNR values of one or greater. This is usually readily achieved in practice by the associated signal conditioning circuits illustrated in the examples that follow. Ideal matched-filter signal conditioning makes use of both amplitude and phase information in upgrading signal quality, and is

Table 4.1 Signal-to-Noise Ratio (SNR) vs. Amplitude and Phase Errors

SNR	Amplitude Error, Random $\varepsilon_{\%FS}$	Phase Error, Random $\varepsilon_{\phi deg}$	Amplitude Error, Coherent $\varepsilon_{\%FS}$
10^1	44.0	22.3	31.1
10^2	14.0	7.07	9.9
10^3	4.4	2.23	3.1
10^4	1.4	0.707	0.990
10^5	0.44	0.223	0.311
10^6	0.14	0.070	0.099
10^7	0.044	0.022	0.0311
10^8	0.014	0.007	0.0099
10^9	0.0044	0.002	0.0031
10^{10}	0.0014	0.0007	0.00099
10^{11}	0.00044	0.0002	0.00031
10^{12}	0.00014	0.00007	0.00009

implied in these SNR relationships for amplitude and phase error in the case of random interference.

For practical applications the SNR requirements ascribed to amplitude and phase error must be mathematically related to conventional amplifier and linear filter signal conditioning circuits. Early work by Fano showed that under high-input SNR conditions, linear filtering approaches matched filtering in its efficiency. Later work by Budai developed a relationship for this efficiency expressed by the characteristic curve of Figure 4.4. This curve and its K parameter appear most reliable for filter numerical input SNR values between about 10 and 100, with an efficiency, K, of 0.9 for SNR values of 200 and greater.

Equations (4.10) through (4.13) describe the relationships upon which the improvement in signal quality may be determined. Both rms and dc voltage

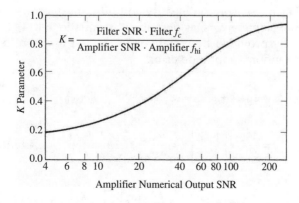

Figure 4.4 Linear filter efficiency, k vs. SNR.

values are interchangeable in equation (4.10). The R_{cm} and R_{diff} impedances of the amplifier input termination account for the V^2/R transducer-gain relationship of the input SNR in equation (4.11). CMRR (common-mode rejection ratio) is squared in this equation in order to convert its ratio of differential to common-mode voltage gains to a dimensionally correct power ratio. Equation (4.12) represents the narrower bandwidth processing-gain relationship for the ratio of amplifier, f_{hi}, to filter, f_c, produced with the filter efficiency, k, for improving signal quality above that provided by the amplifier CMRR with random interference. Most of the improvement is provided by the amplifier CMRR owing to its squared factor, but random noise higher frequency components are also effectively attenuated by linear filtering.

$$\text{Input SNR} = \left(\frac{V_{diff}}{V_{cm}} \right)^2 \text{ dc or rms} \tag{4.10}$$

$$\text{Amplifier SNR} = \text{input SNR} \bullet \frac{R_{cm}}{R_{diff}} \bullet \text{CMRR}^2 \tag{4.11}$$

$$\text{Filter SNR}_{\text{random}} = \text{amplifier SNR} \bullet k \bullet \frac{f_{hi}}{f_c} \tag{4.12}$$

$$\text{Filter SNR}_{\text{coherent}} = \text{amplifier SNR} \bullet \left[1 + \left(\frac{f_{coh}}{f_c} \right)^{2n} \right] \tag{4.13}$$

For coherent interference conditions, signal-quality improvement is a function of achievable filter attenuation at the interfering frequencies. This is expressed by equation (4.13) for one-pole RC to n-pole Butterworth low-pass filters. Note that filter cutoff frequency is determined from the considerations of Tables 3.5 and 3.6 with regard to minimizing the filter error contribution. Finally, the various signal conditioning device errors and output signal quality must be appropriately combined in order to determine total channel error. Sensor nonlinearity and filter mean errors are combined with the RSS of signal errors, as described by equation (4.14). Substitutions are conveniently provided by equations (4.15) and (4.16), respectively, for coherent and random amplitude error.

$$\varepsilon_{\text{channel}} = \bar{\varepsilon}_{\text{sensor}} + \bar{\varepsilon}_{\text{filter}} + \left[\varepsilon_{\text{amplifier}}^2 + \varepsilon_{\text{random}}^2 + \varepsilon_{\text{coherent}}^2 \right]^{1/2} \tag{4.14}$$

$$\varepsilon_{\text{coherent}} = \frac{V_{cm}}{V_{diff}} \bullet \left[\frac{R_{diff}}{R_{cm}} \right]^{1/2} \bullet \frac{\text{Av}_{cm}}{\text{Av}_{diff}} \bullet \left[1 + \left(\frac{f_{coh}}{f_c} \right)^{2n} \right]^{-1/2} \bullet 100\% \tag{4.15}$$

$$\varepsilon_{\text{random}} = \frac{V_{cm}}{V_{diff}} \bullet \left[\frac{R_{diff}}{R_{cm}} \right]^{1/2} \bullet \frac{\text{Av}_{cm}}{\text{Av}_{diff}} \bullet \left[\frac{2}{k} \left(\frac{f_c}{f_{hi}} \right) \right]^{1/2} \bullet 100\% \tag{4.16}$$

4.3 DC, sinusoidal, and harmonic signal conditioning

Signal conditioning is concerned with upgrading the quality of a signal to the accuracy of interest coincident with signal acquisition, scaling, and band-limiting. The unique requirements of each analog data acquisition channel plus the economic constraint of achieving only the performance necessary in specific applications are an impediment to standardized designs. The purpose of this chapter therefore is to develop a unified, quantitative design approach for signal acquisition and conditioning that offers new under-standing and accountability measures. The following examples include both device and system errors in the evaluation of total signal conditioning channel error.

We shall consider a dc and sinusoidal signal conditioning channel that has widespread industrial application in process control and data logging systems. Temperature measurement employing a Type-C thermocouple is to be implemented over the range of 0 to 1800°C while attenuating ground-conductive and electromagnetically coupled interference. A 1-Hz signal bandwidth is coordinated with filter cutoff to minimize the error provided by a single-pole filter as described in Table 3.5. Narrowband signal conditioning is accordingly required for the differential-input 17.2 µV/°C thermocouple signal range of 0 to 31 mV dc, and for rejecting 1 V rms of 60 Hz common-mode interference providing a residual coherent error of 0.009%FS (full scale). An OP-07A subtractor instrumentation ampli-fier circuit combining a 22-Hz differential-lag RC lowpass filter is capable of meeting these requirements, including a full-scale output signal of 4.096 V dc with a differential gain, Av_{diff}, of 132, without the cost of a separate active filter.

This austere dc and sinusoidal circuit is shown by Figure 4.5, with its parameters and defined-error performance tabulated in Tables 4.3 through 4.5. This Av_{diff} further results in a –3-dB frequency response of 4.5 kHz to provide a sensor-loop internal noise contribution of 4.4 μV_{pp} with 100 Ω source resistance. With 1% tolerance resistors the subtractor amplifier presents a common-mode gain of 0.02 by the considerations of Table 2.2. The OP-07A error budget of 0.103%FS is combined with other channel error contributions including a mean filter error of 0.1 %FS and 0.011 %FS linearized thermocou-ple. The total channel error of 0.246%FS at 1σ expressed in Table 4.5 is domi-nated by static mean error, which is an inflexible error to be minimized through-out all instrumentation systems. Postconditioning linearization software achieves a residual deviation from true temperature values of 0.2°C/1800°C, and active cold-junction compensation for ambient temperature is provided by an AD590 sensor attached to the input terminal strip to within 0.5 °C. Note R_i is 10 KΩ.

The information content of instrumentation signals is described by their amplitude variation with time, or through Fourier transformation by signal bandwidth in Hz. Instrumentation signal types are accordingly classified in

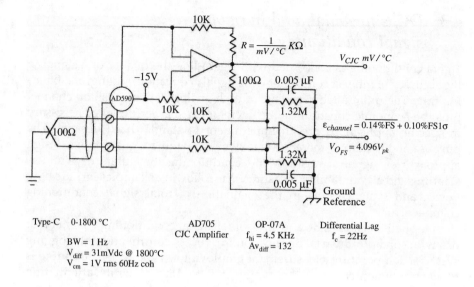

Figure 4.5 DC and sinusoidal signal conditioning.

Table 4.2 with their minimum bandwidth requirements specified in terms of signal waveform parameters. Dc signal time rate of change is equated to the time derivative of a sinusoidal signal and evaluated at zero time to determine its bandwidth requirement. In the case of harmonic signals, a first-order rolloff of -20 dB per decade is assumed from a full-scale signal amplitude at the inverse waveform period, $1/T$, defining the fundamental frequency, declining to one-tenth of full scale at a bandwidth value of ten times the fundamental frequency.

Considered now is the premium harmonic signal conditioning channel of Figure 4.6, employing a 0.1%FS systematic error piezoresistive sensor that can transduce acceleration signals in response to applied mechanical force. Postconditioning signal processing options include subsequent signal integration to obtain velocity and then displacement vibration spectra from these acceleration signals by means of an ac integrator

Table 4.2 Signal Bandwidth (BW)

Signal	BW (Hz)
dc	$dV_s/\pi V_{FS}dt$
Sinusoidal	$1/\text{period } T$
Harmonic	$10/\text{period } T$
Single event	$2/\text{width } \tau$

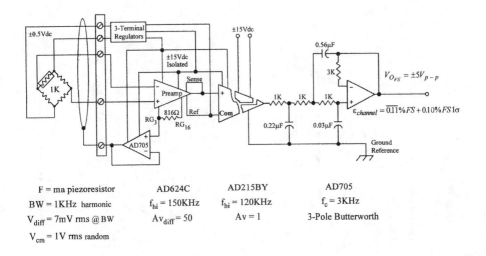

F = ma piezoresistor	AD624C	AD215BY	AD705
BW = 1KHz harmonic	f_{hi} = 150KHz	f_{hi} = 120KHz	f_c = 3KHz
V_{diff} = 7mV rms @ BW	Av_{diff} = 50	Av = 1	3-Pole Butterworth
V_{cm} = 1V rms random			

Figure 4.6 Premium harmonic signal conditioning.

or by digital signal processing. A harmonic signal spectral bandwidth is allowed for this example from dc to 1 KHz with the 1-K-source- resistance bridge sensor generating a maximum input signal amplitude of 70 mV rms, up to 100 Hz fundamental frequencies, with rolloff at –20 dB per decade of frequency to 7 mV rms at 1 KHz bandwidth. The ±0.5-V dc bipolar sensor excitation is furnished by isolated three-terminal regulators to within ±50 µV dc variation, providing a negligible 0.01%FS differential -mode error. The sensor shield buffered common-mode voltage active drive also preserves signal conditioning CMRR over extended cable lengths.

An AD624C preamplifier raises the differential sensor signal to a ±5-V_{pp} full-scale value while attenuating 1 V rms of common-mode random interference, in concert with the lowpass filter, to a residual error of 0.006%FS, as defined by equation (4.16). The error budgets of the preamplifier and isolation amplifier, tabulated in Tables 4.3 and 4.4, also include a sensor -loop internal noise contribution of 15 μV_{pp} based on the provisions of Figure 4.7, where the $1/f$ contact noise frequency is taken as 10% of signal bandwidth. Three contributions comprising this internal noise are evaluated as source resistance thermal noise, V_t, contact noise, V_c, arising from 1 mA of DC current flow, and amplifier internal noise, V_n. The three-pole Butterworth lowpass filter cutoff frequency is derated to a value of 3 BW to minimize its device error. Note that the AD705 filter amplifier is included in the mean filter device error of 0.115 %FS. The total channel 1σ instrumentation error of 0.221%FS consists of an approximate equal sum of mean and systematic error values at 1σ confidence in Table 4.5.

DC & Sinusoidal Channel   Harmonic Channel

Sensor

Type-C thermocouple 17.2 μV/°C
post-conditioning linearization

$$\text{software } \frac{\overline{0.2°C}}{1800°C} \bullet 100\% = \overline{0.011}\ \%\text{FS}$$

1 KΩ piezoresistor bridge
with F = ma response

0.1%FS

Interface

AD 590 temperature sensor
cold-junction compensation

$$\frac{\overline{0.5°C}}{1800°C} \bullet 100\% = \overline{0.032}\%\text{FS}$$

Regulators for sensor excitation

±0.5 V DC ± 50 μV or 0.01%FS

Signal Quality

$$\varepsilon_{coh} = \frac{V_{cm}}{V_{diff}} \bullet \left[\frac{R_{diff}}{R_{cm}}\right]^{1/2} \bullet \frac{Av_{cm}}{Av_{diff}} \bullet \left[1 + \left(\frac{f_{coh}}{f_c}\right)^{2n}\right]^{-1/2} \bullet 100\%$$

$$= \frac{\left(1\,V_{rms}\,2\sqrt{2}\,\right)_{pp}}{31\,mV_{dc}} \bullet \left[\frac{80\,M\Omega}{200\,G\Omega}\right]^{1/2} \bullet \frac{0.02}{132} \bullet \left[1 + \left(\frac{60\,Hz}{22\,Hz}\right)^2\right]^{-1/2} \bullet 100\%$$

$$= 0.009\%\text{FS}$$

$$\varepsilon_{rand} = \frac{V_{cm}}{V_{diff}} \bullet \left[\frac{R_{diff}}{R_{cm}}\right]^{1/2} \bullet \frac{Av_{cm}}{Av_{diff}} \bullet \left[\frac{2}{K}\frac{f_c}{f_{hi}}\right]^{1/2} \bullet 100\%$$

$$= \frac{1\,V}{7\,mV} \bullet \left[\frac{1\,G\Omega}{1\,G\Omega}\right]^{1/2} \bullet \frac{10^{-4}}{50} \bullet \left[\frac{2}{0.9}\frac{3\,kHz}{150\,kHz}\right]^{1/2} \bullet 100\%$$

$$= 0.006\%\text{FS}$$

Table 4.3 Amplifier Parameter Values

Symbol	OP-07A	AD624C	AD215BY	Comment
V_{OS}	10 μV	25 μV	0.4 mV	Offset voltage
$\dfrac{dV_{OS}}{dT}$	0.2 μV/°C	0.25 μV/°C	2 μV/°C	Voltage drift
I_{OS}	0.3 nA	10 nA	300 nA	Offset current
$\dfrac{dI_{OS}}{dT}$	5 pA/°C	20 pA/°C	1 nA/°C	Current drift
Av_{diff}	132	50	1	Differential gain
Av_{cm}	0.02 (1%R)	0.0001	0.0001	Common-mode gain
CMRR	6600	5×10^5	10^4	Av_{diff}/Av_{cm}
CMV	10 V_{rms}	10 V_{rms}	1500 V_{rms}	Max. applied volts
$V_{N_{pp}}$	$6.6\left[(V_t^2+V_n^2)f_{hi}\right]^{1/2}$	$6.6\left[(V_t^2+V_c^2+V_n^2)f_{hi}\right]^{1/2}$	$6.6\left[V_t^2 f_{hi}\right]^{1/2}$	Total input noise
V_t rms	1.3 nV/√Hz	4 nV/√Hz	0.9 nV/√Hz	Thermal noise
V_c rms	None	1.8 nV/√HZ	Negligible	Contact noise
V_n rms	10 nV/√HZ	4 nV/√HZ	Negligible	Amplifier noise
f_{hi}	4.5 KHz	150 KHz	120 KHz	–3-dB bandwidth
$f_{contact}$	None	100 Hz	100 Hz	Contact noise frequency
$\dfrac{dA_v}{dT}$	50 ppm/°C	5 ppm/°C	15 ppm/°C	Gain drift
$f(Av)$	0.01%	0.001%	0.005%	Gain nonlinearity
R_{diff}	$8 \times 10^7\ \Omega$	$10^9\ \Omega$	$10^{12}\ \Omega$	Differential resistance
R_{cm}	$2 \times 10^{11}\ \Omega$	$10^9\ \Omega$	$5 \times 10^9\ \Omega$	Common-mode resistance
R_s	100 Ω	1 K	50 Ω	Source resistance
V_{OPS}	4.096 V_{pk}	±5 V_{pp}	±5 V_{pp}	Full-scale output
$\dfrac{dV_{OPS}}{dT}$	10°C	10°C	10°C	Temperature variation

Table 4.4 Amplifier Error Budgets

$\varepsilon_{ampl_{RTI}}$	OP-07A	AD624C	AD215BY
V_{os}	$\overline{10}$ μV	Trimmed	Trimmed
$\dfrac{dV_{os}}{dT} \bullet dT$	2 μV	2.5 μV	20 μV
$I_{os} \bullet R_i$	$\overline{3}$ μV	$\overline{10}$ μV	$\overline{15}$ μV
$V_{N_{pp}}$	4.4 μV	15 μV	2 μV
$f(Av) \bullet \dfrac{V_{O_{FS}}}{Av_{diff}}$	$\overline{3}$ μV	$\overline{1}$ μV	$\overline{250}$ μV
$\dfrac{dA_V}{dT} \bullet dT \bullet \dfrac{V_{O_{FS}}}{Av_{diff}}$	15.5 μV	5 μV	750 μV
$\Sigma\ \overline{mean} + 1\sigma$ RSS	$(\overline{16}+16)$μV	$(\overline{11}+16)$μV	$(\overline{265}+750)$μV
$X\ \dfrac{Av_{diff}}{V_{O_{FS}}} \bullet 100\%$	0.103%FS	0.027%FS	0.020%FS

Table 4.5 Signal Conditioning Channel Error Summary

Element	DC & Sinusoidal $\varepsilon_{\%FS}$	Comment	Harmonic $\varepsilon_{\%FS}$	Comment
Sensor	$\overline{0.011}$	Type C linearized	0.100	Piezoresistor
Interface	$\overline{0.032}$	CJC sensor	0.010	Sensor excitation
Amplifier	0.103	OP-07A	0.027	AD624C
Isolator	None		0.020	AD215AY
Filter	$\overline{0.100}$	Table 3-5	0.115	Table 3-6
Signal quality	0.009	60 Hz ε_{coh}	0.006	Noise ε_{rand}
$\varepsilon_{channel}$	0.143 %FS	$\Sigma\ \overline{mean}$	0.115 %FS	$\Sigma\ \overline{mean}$
	0.103%FS	1σ RSS	0.106%FS	1σ RSS
	0.246%FS	$\Sigma\ \overline{mean}+1\sigma$RSS	0.221%FS	$\Sigma\ \overline{mean}+1\sigma$RSS

4.4   Analog signal processing

When achievable analog signal conditioning error does not meet minimum measurement requirements identical channels may be averaged to reduce the total error. Random and systematic errors added to the value of a measurement can be reduced by taking the arithmetic mean of a sum of n independent measurement values. This assumes that combined systematic error contributions are sufficient in number to approximate a zero mean value, and as well for random errors. Sensor device error is frequently simplified in its specification as the nonlinearity of its transfer function and conservatively represented by a mean error. However, many effects actually contribute to sensor error, such as material–energy interactions, which are unknown other than their dependence on random variables that generally are compliant to reduction by arithmetic-mean averaging.

The foregoing conditions are sufficiently met by typical signal conditioning channels to enable averaged outputs consisting of arithmetic signal additions and RSS error additions. This provides signal quality improvement by $n/\sqrt{n}$ and channel error reduction by its inverse. Averaged measurement error accordingly corresponds to the error of any one identical channel divided by $\sqrt{n}$. However, diminishing returns may result in an economic penalty to achieve error reduction beyond a few channels combined. Further, signal conditioning mean filter device error also remains additive, which is a limitation remedied by relocating the channel filter after averaging.

Figure 4.7 describes signal conditioning channel averaging where amplifier stacking between respective device outputs and ground provide arithmetic signal additions, and their parallel inputs provide RSS error additions. The Av_{diff} values of each stage are equally scaled so that the sum of n outputs achieves the full-scale value for a single channel. The three averaged harmonic signal conditioning channels, therefore, each require an Av_{diff} of 16.67 for a per-channel output of 1.667 V by employing gain resistors of 2552 Ω. With reference to Table 4.5, moving the filter postaveraging provides an improved overall error of $(0.221\% - \overline{0.115\%})/\sqrt{3} + 0.115\%$, approximately totaling $\overline{0.176\%}$FS. Note that this connection obviates the requirement for an output summing amplifier and its additional device error contribution.

The utility of logarithmic functions in signal processing primarily lies with their ability to accommodate wide-dynamic-range signals. Logarithmic arguments are always dimensionless. Consequently, the log of the ratio of two voltages or currents is required in logarithmic realizations, with the denominator term normally a fixed reference value, as described in Table 4.6. Commercially available log devices offer performance extending over four voltage decades, typically from 1 mV to 10 V, and six current decades from 1 nA to 1 mA. The bipolar log function is especially useful and is symmetrical about its linear segment centered on zero input signals, shown in Figure 4.8. Present analog circuits utilize a silicon bipolar transistor to form the logarithmic function with its forward-biased base-emitter junction.

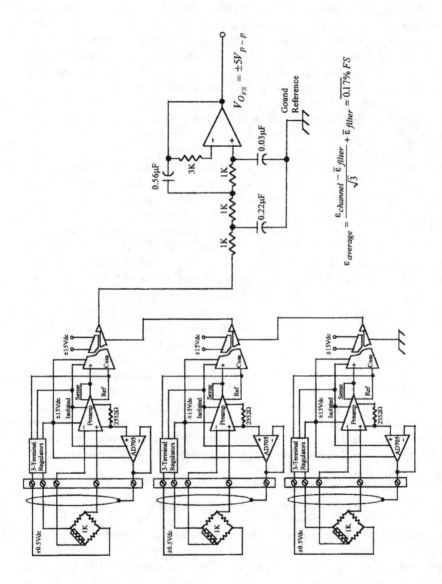

Figure 4.7 Signal conditiong error averaging.

Table 4.6 Logarithmic Functions

Function	Description
$V_o = C \cdot \log_{10}\left(\dfrac{V_i}{V_{ref}}\right)$	Voltage log
$V_o = C \cdot \log_{10}\left(\dfrac{I_i}{I_{ref}}\right)$	Current log
$V_o = V{ref} \cdot 10^{-V_i/C}$	Antilog
$V_o = C \cdot \log_{10}\left(\dfrac{V_1}{V_2}\right)$	Log ratio
$V_o = C \cdot \sinh^{-1}\left(\dfrac{V_i}{2V_{ref}}\right)$	Bipolar log
$V_o = C \cdot \sinh\left(\dfrac{V_i}{V_{ref}}\right)$	Bipolar antilog

The application of a logarithmic function at the input of a data acquisition system has merit for processing high-resolution sensor data by compressing an input signal to conform to limited system dynamic range. This can be appreciated from observation of Figure 4.8, where it is apparent that the logarithmic gain is unity for a 10-V full-scale input signal, and logarithmically increases to a maximum gain at minimum input levels. However, a mirror-symmetry antilog operation is required following output signal recovery for linear signal representation. This input compression and output expansion, or companding operation, is beneficial in extending the dynamic

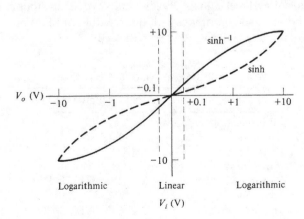

Figure 4.8 Bipolar log and antilog functions.

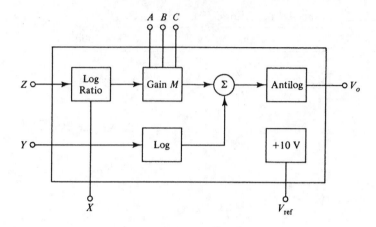

Figure 4.9 Log–antilog multifunction module.

range of truncated-wordlength digital processors by effectively increasing the number of A/D converter quantizing levels at lower signal amplitudes. The typical log conformity error is 1%FS, and this constant fractional error is maintained throughout the signal dynamic range in logarithmic form at the expense of high signal resolution at any point within the range.

The requirement occasionally arises in instrumentation and process control systems to perform specific computations, such as an rms extraction or mass flow calculation, where a digital implementation with a microcomputer including the required data conversion is not economical. The logarithmic multifunction module described by Figure 4.9 combines a number of mathematical functions within the device, including signal multiplication, division, and the ability to raise a ratio of signals to a power or a root. Examples of these operations can be found in Table 4.7.

An occasionally required signal processing operation is the precision rectification of very-low-level signals. Passive fullwave rectification is inadequate for this task because silicon diodes will not conduct until an applied voltage exceeds approximately 600 mV. However, the active fullwave rectification and

Table 4.7 Multifunction Module Operations

Function	Transfer Equation
Multiplication	$YZ/10$
Division	$10(Z/X)$
Root of ratio	$Y(Z/X)^M \ M < 1$
Power of ratio	$Y(Z/X)^M \ M > 1$
Reciprocal power	$Y(Z/X)^M = Y(X/Z)^{-M}$
True rms	$\overline{V_i^2}/V_o$
Trigonometric	$\sin Z = Z - 0.17(Z)^{2.83}$
	$\cos Z = 1 + 0.235Z - 0.7(Z)^{1.5}$

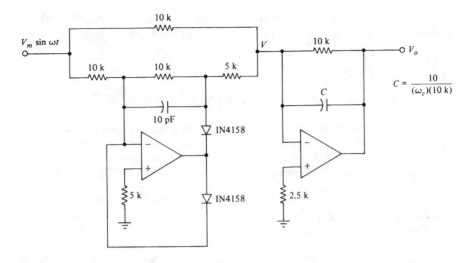

Figure 4.10 Precision AC-to-DC converter.

smoothing circuit of Figure 4.10 will provide accurate AC-to-DC conversion down to submillivolt signal levels. The circuit shown generates a conversion error of 0.6%FS, determined primarily by the residual harmonic distortion passed by the RC smoothing filter with a cutoff frequency of one-tenth the input signal frequency.

AC integrators are useful for providing acceleration-to-velocity or velocity -to-displacement transformations and vice versa, depending on which slope of the response curve the signal frequencies are applied to. Accurate integration to very low frequencies is possible with the circuit of Figure 4.11 without stability problems owing to the attenuation provided by the $1/R_1C_1$ break frequency. The symmetrical tee network of this circuit also enables

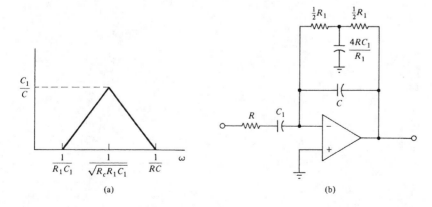

Figure 4.11 AC integrator: (a) response, (b) circuit.

extended-frequency differentiation, while maintaining noise rejection at still higher frequencies determined by the position of the $1/RC$ break frequency.

Bibliography

1. Akao, Y., "Quality Function Deployment and CWQC in Japan," *Quality Progress*, 1983.
2. Budai, M., "Optimization of the Signal Conditional Channel," Senior Design Project, *Electrical Engineering Technology*, University of Cincinnati, 1978.
3. *Designers Reference Manual*, Analog Devices, Norwood, MA, 1996.
4. Fano, R.M., "Signal to Noise Ratio in Correlation Detectors," *MIT Technical Report 186*, 1951.
5. Garrett, P.H., *Analog Systems for Microprocessors and Minicomputers*, Reston Publishing Co., Reston, VA, 1978.
6. Garrett, P.H., *Advanced Instrumentation and Computer I/O Design*, IEEE Press, New York, 1994.
7. Gordon, B.M., *The Analogic Data-Conversion Systems Digest*, 2nd edition Analogic, Wakefield, MA, 1977.
8. Petriu, E.M., Ed., *Instrumentation and Measurement Technology and Applications*, IEEE Technology Update Series, New York, 1998.
9. Raemer, H.R., *Statistical Communications Theory and Applications*, Prentice Hall, New York, 1969.
10. Schwartz, M., Bennett, W., and Stein, S., *Communications Systems and Techniques*, McGraw-Hill, New York, 1966.
11. Sheingold, D.H., Ed., *Transducer Interfacing Handbook*, Analog Devices, Norwood, MA, 1980.
12. Ott, H.W., *Noise Reduction Techniques in Electronic Systems*, Wiley Interscience, New York, 1976.

chapter five

Data conversion devices and errors

5.0 Introduction

Data conversion devices provide the interfacing components between continuous-time signals representing the parameters of physical processes and their discrete-time digital equivalent. Recent emphasis on computer systems for automated manufacturing and the growing interest in using personal computers for data acquisition and control have increased the need for improved understanding of the design requirements for real-time I/O systems. However, before describing the theory and practice involved in these systems it is advantageous to understand the characterization and operation of the specific devices from which these systems are fabricated. This chapter provides design information concerning analog-to-digital (A/D) and digital-to-analog (D/A) data conversion devices, including seven application-specific A/D converters, with supporting components including analog multiplexers and sample-hold devices. The development of the individual error budgets representing these devices is also provided to continue the quantitative methodology of this text.

5.1 Analog multiplexers

Field-effect transistors, both CMOS and JFET, are universally used as electronic multiplexer switches, displacing earlier bipolar devices because of their voltage offset problems. Junction FET (JFET) switches have greater device electrical ruggedness and approximately the same switching speeds as CMOS devices. However, CMOS switches are dominant in multiplexer applications because of their unfailing turnoff, especially when the power is removed (unlike JFET devices), and their ability to multiplex signal levels up to the power supply voltages. Figure 5.1 describes a CMOS analog switch circuit that achieves a stable ON resistance of about 100 Ω series resistance by the parallel p- and n-channel devices. Terminating a CMOS multiplexer with a high-input-impedance voltage

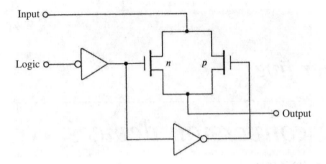

Figure 5.1 CMOS analog switch.

follower eliminates voltage-divider errors possible as a consequence of the ON resistance. Figure 5.2 presents interconnection configurations for a multiplexer. These switches are characterized by Table 5.1.

Errors associated with analog multiplexers are tabulated in Table 5.2 and are dominated by the average transfer error defined by equation (5.1). This error is essentially determined by the input voltage-divider effect and is minimized to a typical value of 0.01%FS when the AMUX is followed by an output buffer amplifier. The input amplifier associated with a sample-hold

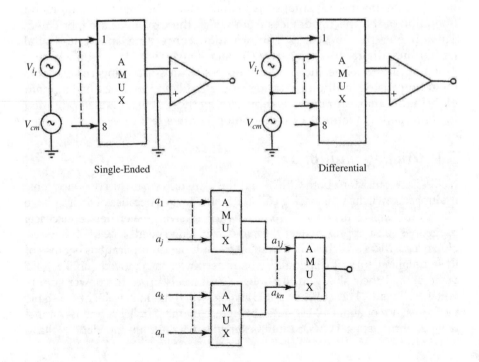

Figure 5.2 Multiplexer interconnections and array.

Table 5.1 Multiplexer Switch Characteristics

Type	ON Resistance	OFF Isolation	Sample Rate
CMOS	100 Ω	70 dB	10 MHz
JFET	50 Ω	70 dB	1 MHz
Reed	0.1 Ω	90 dB	1 kHz

Table 5.2 Representative Multiplexer Errors

Parameter		Reed	CMOS
Transfer error		0.01%	0.01%
Crosstalk error		0.001	0.001
Leakage error			0.001
Thermal offset		0.001	
ε_{AMUX}	$\Sigma \overline{mean} + 1\sigma$ RSS	0.01%FS	0.01%FS

device often provides this high-impedance termination. Another error that can be significant is OFF-channel leakage current that creates an offset voltage across the input source resistance.

$$\text{Transfer error} = \frac{V_i - V_o}{V_i} \times 100\% \qquad (5.1)$$

5.2 Sample-holds

Sample-hold (S/H) devices provide an analog signal memory function to be used in sampled-data systems for temporary storage of changing signals for data conversion purposes. Sample-holds are available in several circuit variations, each suited to specific speed and accuracy requirements. Figure 5.3 describes a contemporary circuit that may be optimized for either speed or accuracy. The noninverting input amplifier provides a high-impedance buffer stage, and the overall unity feedback minimizes signal transfer error

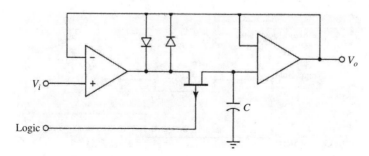

Figure 5.3 Closed-loop sample-hold.

Table 5.3 Representative Sample-Hold Errors

Acquisition error	0.01%
Nonlinearity	0.004
Gain	0.01
Temperature coefficient	0.001
$\varepsilon_{S/H}$ $\sum \overline{mean} + 1\sigma$ RSS	0.02%FS

when the device is in the tracking mode. The clamping diodes ensure that the circuit remains stable during the hold mode when the switch is open. The inclusion of S/H devices in sampled-data systems must be carefully considered. The following examples represent the three essential applications for sample-holds.

Figure 5.4 diagrams a conventional multiplexed data conversion system cycle. The multiplexer and external circuit of Channel 1 are sampled by the sample-hold for a time sufficient for signal settling to within the amplitude error of interest. For sensor channels having RC time constants on the order of the S/H internal acquisition time, defined by equation (5.2), overlapping multiplexer channel selection and A/D conversion can speed system throughput significantly by means of an interposed sample-hold. A second application is described by Figure 5.5. Simultaneous data acquisition is required for many measurements for which multiple sensor channels must be acquired at precisely the same time. Matching S/H devices in bandwidth and aperture time can minimize interchannel signal time skew. The timing relationships are consequently preserved between signals even though data conversion is performed sequentially.

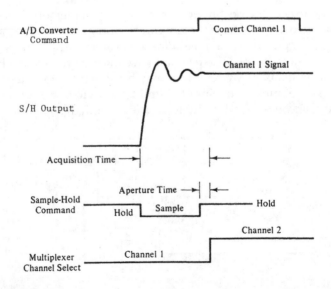

Figure 5.4 Multiplexed conversion system timing diagram.

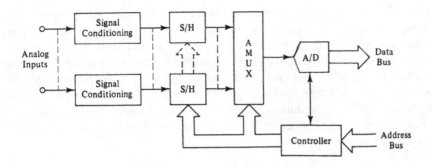

Figure 5.5 Simultaneous data acquisition.

Voltage-comparison A/D converters such as successive approximation devices require a constant signal value for accurate conversion. This function is normally provided by the application of a sample-hold preceding the A/D converter, which constitutes the third application. An important issue is matching of S/H and A/D specifications to achieve the performance of interest. Sample-hold performance is principally determined by the input amplifier bandwidth and current output capability, which determines its ability to drive the hold capacitor, C. A limiting parameter is the acquisition time of equation (5.2) and Figure 5.6, which when added to the conversion period, T, of an A/D converter determines the maximum throughput performance possible for a sample-hold and connected A/D. As a specific example, an Analog Devices 9100 device has an acquisition time of 14 nsec for 0.01%FS (13-bit) settling, enabling data conversion rates $\frac{1}{14ns + T}$ Hz. In the sample mode the charge on

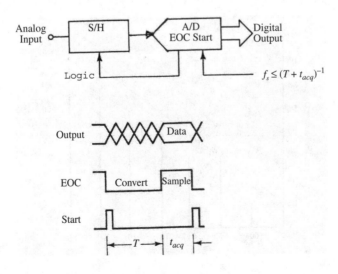

Figure 5.6 S/H–A/D timing relationships.

the hold capacitor is initially changed at the slew-limited output current capability, I_o, of the input amplifier. As the capacitor voltage enters the settling band coincident with the linear region of amplifier operation, final charging is exponential and corresponds to the summed time constants in equation (5.2), where R_o corresponds to amplifier output resistance and R_{ON} the switch resistance. The consequence of aperture time is to provide an average aperture error associated with the finite bound within which the amplitude of a sampled signal is acquired. Because this is a system error instead of a component error its evaluation is deferred to Section 6.3.

$$\text{Acquisition time} = \frac{|V_o - V_i|}{I_o} + 9(R_o + R_{ON})C \text{ sec} \qquad (5.2)$$

5.3   Digital-to-analog converters

D/A converters, or DACs, provide reconstruction of discrete-time digital signals into continuous-time analog signals for computer interfacing output data recovery purposes such as actuators, displays, and signal synthesizers. DACs are considered prior to A/D converters because some A/D circuits require DACs in their implementation. A D/A converter may be considered a digitally controlled potentiometer that provides an output voltage or current normalized to a full-scale reference value. A descriptive way of indicating the relationship between analog and digital conversion quantities is a graphical representation. Figure 5.7 describes a 3-bit D/A converter transfer relationship with eight analog output levels ranging from 0 to $\frac{7}{8}$ of full scale. Notice that a DAC full-scale digital input code produces an analog output

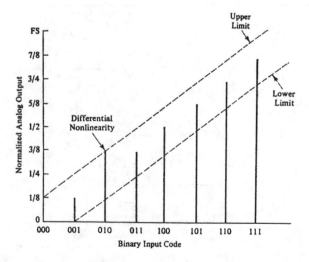

Figure 5.7 Three-bit D/A converter relationships.

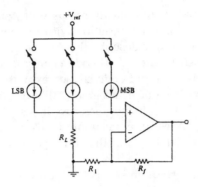

Figure 5.8 Straight binary 3-bit DAC.

equivalent to FS – 1 LSB (least significant bit). The basic structure of a conventional D/A converter includes a network of switched current sources having MSB (most significant bit) to LSB values according to the resolution to be presented. Each switch closure adds a binary-weighted current increment to the output bus. These current contributions are then summed by a current-to-voltage converter amplifier in a manner appropriate to scale the output signal. Figure 5.8 illustrates such a structure for a 3-bit DAC, with unipolar straight binary coding corresponding to the representation of Figure 5.7.

In practice, the realization of the transfer characteristic of a D/A converter is nonideal. With reference to Figure 5.7, the zero output may be nonzero because of amplifier offset errors, the total output range from 0 to FS – 1 LSB may have an overall increasing or decreasing departure from the true encoded values resulting from gain error, and differences in the height of the output bars may exhibit a curvature owing to converter nonlinearity. Gain and offset errors may be compensated for by leaving the residual temperature-drift variations shown in Table 5.4 as the temperature coefficients of a representative 12-bit DAC. A voltage reference is also necessary to establish a basis for the DAC absolute output voltage. The majority of voltage references utilize the bandgap principle, whereby the V_{be} of a silicon transistor has a negative temperature coefficient of –2 mV/°C that can be extrapolated to approximately 1.2 V at absolute zero (the bandgap voltage of silicon).

Converter nonlinearity is minimized through precision components, because nonlinearity is essentially distributed throughout the converter network

Table 5.4 Representative 12-Bit DAC Errors

Mean integral nonlinearity (1 LSB)		0.024%
Temperature coefficient (1 LSB)		0.024
Noise + distortion		0.001
$\varepsilon_{D/A}$	$\sum mean + 1\sigma$ RSS	0.048%FS

and cannot be eliminated by adjustment, as with gain and offset errors. Differential nonlinearity and its variation with temperature are prominent in data converters in that they describe the difference between the true and actual outputs for each of the 1-LSB code changes. A DAC with a 2-LSB output change for a 1-LSB input code change exhibits 1 LSB of differential nonlinearity as shown. Nonlinearities greater than 1 LSB make the converter output no longer single valued, in which case it is said to be nonmonotonic and to have missing codes. Integral nonlinearity is an average error that generally does not exceed 1 LSB of the converter resolution as the sum of differential nonlinearities.

Table 5.5 presents frequently applied unipolar and bipolar codes expressed in terms of a 12-bit binary wordlength. These codes are applicable to both D/A and A/D converters. Code choice should be appropriate to the application and its sense understood (positive-true, negative-true). Positive-true coding defines a logic 1 as the positive logic level, and in negative-true coding the negative logic level is 1 with the other level 0. All codes utilized with data converters are based on the binary number system. Any base 10 number may be represented by equation (5.3), where the coefficient a_i assumes a value of 1 or 0 between the MSB (0.5) and LSB (2^{-n}).

$$N = \sum_{i=0}^{n} a_i \, 2^{-i} \qquad (5.3)$$

This coding scheme is convenient for data converters where the encoded value is interpreted in terms of a fraction of full scale for *n*-bit lengths. Straight-binary, positive-true unipolar coding is most commonly encountered. Complementary binary positive-true coding is identical to straight binary negative-true coding. Sign-magnitude bipolar coding is often used for outputs that are frequently in the vicinity of zero. Offset binary is readily converted to the more computer-compatible twos-complement code by complementing the MSB.

As the input code to a DAC is increased or decreased it passes through major and minor transitions. A major transition is at half-scale when the MSB is switched and all other switches change state. If some switched current sources lag others, then significant transient spikes are generated, known as glitches. Glitch energy is of concern in fast-switching DACs driven by high-speed logic with time skew between transitions. However, high-speed DACs also frequently employ an output S/H circuit to deglitch major transitions by remaining in the hold mode during these intervals. Internally generated noise is usually not significant in DACs except at extreme resolutions, such as the 20-bit Analog Devices DAC 1862, the LSB of which is equal to 10 μV with 10-V_{FS} scaling.

The advent of monolithic DACs has resulted in almost universal acceptance of the R-2R network DAC because of the relative ease of achieving

Table 5.5 Data Converter Binary Codes

	Unipolar Codes — 12-Bit Converters Straight Binary and Complementary Binary			
Scale	+10 V$_{FS}$	+5 V$_{FS}$	Straight Binary	Complementary Binary
+FS – 1 LSB	+9.9976	+4.9988	1111 1111 1111	0000 0000 0000
+7/8 FS	+8.7500	+4.3750	1110 0000 0000	0001 1111 1111
+3/4 FS	+7.5000	+3.7500	1100 0000 0000	0011 1111 1111
+5/8 FS	+6.2500	+3.1250	1010 0000 0000	0101 1111 1111
+1/2 FS	+5.0000	+2.5000	1000 0000 0000	0111 1111 1111
+3/8 FS	+3.7500	+1.8750	0110 0000 0000	1001 1111 1111
+1/4 FS	+2.5000	+1.2500	0100 0000 0000	1011 1111 1111
+1/8 FS	+1.2500	+0.6250	0010 0000 0000	1101 1111 1111
0+1 LSB	+0.0024	+0.0012	0000 0000 0001	1111 1111 1110
0	0.0000	0.0000	0000 0000 0000	1111 1111 1111

	BCD and Complementary BCD			
Scale	+10 V$_{FS}$	+5 V$_{FS}$	Straight Binary	Complementary Binary
+FS – 1 LSB	+9.99	+4.95	1001 1001 1001	0110 0110 0110
+7/8 FS	+8.75	+4.37	1000 0111 0101	0111 1000 1010
+3/4 FS	+7.50	+3.75	0111 0101 0000	1000′ 1010 1111
+5/8 FS	+6.25	+3.12	0110 0010 0101	1001 1101 1010
+1/2 FS	+5.00	+2.50	0101 0000 0000	1010 1111 1111
+3/8 FS	+3.75	+1.87	0011 0111 0101	1100 1000 1010
+1/4 FS	+2.50	+1.25	0010 0101 0000	1101 1010 1111
+1/8 FS	+1.01	+0.62	0001 0010 0101	1110 1101 1010
0+1 LSB	+0.00	+0.00	0000 0000 0001	1111 1111 1110
0	0.00	0.00	0000 0000 0000	1111 1111 1111

	Bipolar Codes — 12-Bit Converters			
Scale	±5 V$_{FS}$	Offset Binary	Twos Complement	Ones Complement	Sign-Mag Binary
+ FS – 1 LSB	4.9976	1111 1111 1111	0111 1111 1111	0111 1111 1111	1111 1111 1111
+3/4 FS	+3.7500	1110 0000 0000	0110 0000 0000	0110 0000 0000	1110 0000 0000
+1/2 FS	+2.5000	1100 0000 0000	0100 0000 0000	0100 0000 0000	1100 0000 0000
+1/4 FS	+1.2500	1010 0000 0000	0010 0000 0000	0010 0000 0000	1010 0000 0000
0	0.0000	1000 0000 0000	0000 0000 0000	0000 0000 0000	1000 0000 0000
–1/4 FS	–1.2500	0110 0000 0000	1110 0000 0000	1101 1111 1111	0010 0000 0000
–1/2 FS	–2.5000	0100 0000 0000	1100 0000 0000	1011 1111 1111	0100 0000 0000
–3/4 FS	–3.7500	0010 0000 0000	1010 0000 0000	1001 1111 1111	0110 0000 0000
–FS + 1 LSB	–4.9976	0000 0000 0001	1000 0000 0001	1000 0000 0000	0111 1111 1111
–FS	–5.0000	0000 0000 0000	1000 0000 0000		

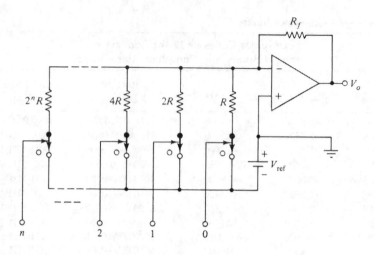

Figure 5.9 Weighted resistor D/A converter.

precise resistance ratios with monolithic technology. This is in contrast to the low yields experienced with achieving precise absolute resistance values required by weighted-resistor networks. Equations (5.4) and (5.5) define the quantities of each converter. For the R-2R network, an effective resistance of 3R is seen by V_{ref} for each branch connection, with equal left–right current division. A weighted resistor DAC circuit is shown by Figure 5.9.

$$V_o = \frac{R_f}{R} \bullet V_{ref} \bullet \sum_{i=0}^{n} 2^{-i} \quad \text{Weighted resistor} \tag{5.4}$$

$$V_o = \frac{R_f}{2R} \bullet \frac{V_{ref}}{3} \bullet \sum_{i=0}^{n} 2^{-i} \quad R - 2R \text{ resistor} \tag{5.5}$$

A DAC that accepts a variable reference can be configured as a multiplying DAC useful for many applications requiring a digitally controlled scale factor. Both linear and logarithmic scale factors are available for applications such as, respectively, digital excitation in test systems and a decibel step attenuator in communications systems. The simplest devices operate in one quadrant with a unipolar reference signal and digital code. Two-quadrant multiplying DACs utilize either bipolar reference signals or bipolar digital codes. Four-quadrant multiplication involves both a bipolar reference signal and bipolar digital code. Table 5.6 describes a two-quadrant, 12-bit linear multiplying DAC. The variable transconductance property made possible by multiplication is useful for many signal conditioning applications including programmable gain.

As system peripheral complexity has expanded to require more of a host computer's resources, peripheral interface devices have included transparent processing capabilities to more efficiently distribute these tasks. In fact,

Table 5.6 Two-Quadrant Multiplying 12-Bit DAC

Straight Binary Input	Analog Output
1111 1111 1111	$\pm V_i \left(\dfrac{4095}{4096} \right)$
1000 0000 0001	$\pm V_i \left(\dfrac{2048}{4096} \right)$
0000 0000 0001	$\pm V_i \left(\dfrac{1}{4096} \right)$
0000 0000 0000	OV

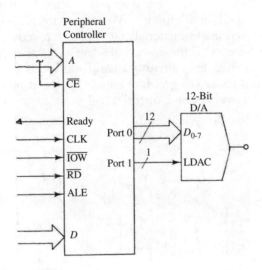

Figure 5.10 D/A peripheral interface.

some devices are more complicated than the host computer they support, such as video and graphics processors. Universal peripheral bus master devices have evolved that offer a flexible combination of memory-mapped, interrupt-driven, and DMA data transfer capabilities with FIFO buffer memory for accommodation of multiple buses and differing speeds. The D/A peripheral interface of Figure 5.10 employs a program-initiated output whose status is polled by the host for a Ready enable. Data may then be transferred to the D port with $\overline{IOW}$ low and $\overline{CE}$ high.

5.4   Analog-to-digital converters

The conversion of continuous-time analog signals to discrete-time digital signals is fundamental to obtaining a representative set of numbers that can be used by a digital computer. The three functions of sampling, quantizing,

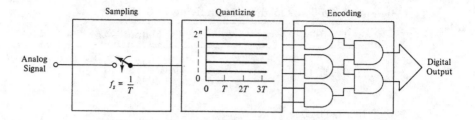

Figure 5.11 A/D converter functions.

and encoding are involved in this process and implemented by all A/D converters, as illustrated by Figure 5.11. The detailed analytical considerations associated with these functions and their relationship to computer interface design are developed in Chapter 6. We are concerned here with A/D converter devices and their functional operations. In practice, one conversion is performed each period T, the inverse of sample rate f_s, whereby a numerical value derived from the converter quantizing levels is translated to an appropriate output code. The graph of Figure 5.12 describes A/D converter input–output relationships and quantization error for prevailing uniform

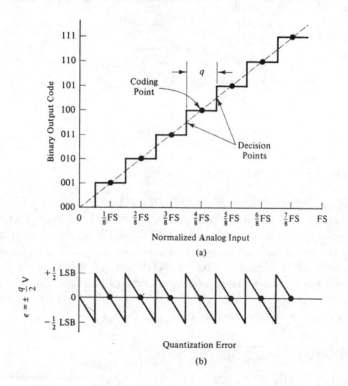

Figure 5.12 Three-bit A/D converter relationships: (a) quantization intervals, (b) quantization error.

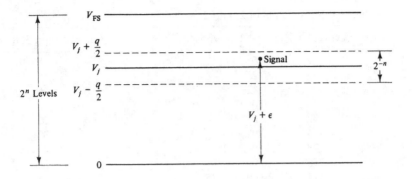

Figure 5.13 Quantization level parameters.

quantization, where each of the levels q is of spacing 2^{-n} (1 LSB) for a converter having an n-bit binary output wordlength. Note that the maximum output code does not correspond to a full-scale input value but instead to $(1 - 2^{-n}) \bullet$ FS, because there exist only $(2^n - 1)$ coding points, as shown in Figure 5.12.

Quantization of a sampled analog waveform involves the assignment of a finite number of amplitude levels corresponding to discrete values of input signal, V_S, between 0 and V_{FS}. The uniformly spaced quantization intervals, 2^{-n}, represent the resolution limit for an n-bit converter, which may also be expressed as the quantizing interval q equal to $V_{FS}/(2^n - 1)$. Figure 5.13 illustrates the prevailing uniform quantizing algorithm whereby an input signal that falls within the V_jth-level range of $\pm \frac{q}{2}$ is encoded at the V_jth level with a quantization error of ε volts. This error may range up to $\pm \frac{q}{2}$ and is an irreducible noise added to a converter output signal. The conventional assumption concerning the probability density function of this noise is that it is uniformly distributed along the interval $\pm \frac{q}{2}$ and is represented as the A/D converter quantizing uncertainty error of value $\frac{1}{2}$ LSB proportional to converter wordlength.

The equivalent rms error of quantization (Eqe) produced by this noise is described by equation (5.6). The rms sinusoidal signal-to-noise ratio (SNR) of equation (5.7) then defines the output signal quality achievable, expressed in power dB, for an A/D converter of n bits with a noise-free input signal. These relationships are tabulated in Table 5.7. Equation (5.8) defines the dynamic range of a data converter of n bits in voltage dB. Converter dynamic range is useful for matching A/D converter wordlength in bits to a required analog input signal span to be represented digitally. For example, a 10-mV-to-10-V span (60 voltage dB) would require a minimum converter wordlength n of 10 bits. It will be shown in Section 6.3 that additional considerations are involved in the conversion of an input signal to an n-bit accuracy, other than the choice of A/D converter wordlength, where the dynamic range of a digitized signal may be represented to n bits without achieving n-bit data accuracy. However, the choice of a long wordlength A/D

Table 5.7 Decimal Equivalents of 2^n and 2^{-n}

Bits, n	Levels, 2^n	LSB Weight, 2^{-n}	Quantization SNR, dB
1	2	0.5	8
2	4	0.25	14
3	8	0.125	20
4	16	0.0625	26
5	32	0.03125	32
6	64	0.015625	38
7	128	0.0078125	44
8	256	0.00390625	50
9	512	0.001953125	56
10	1,024	0.0009765625	62
11	2,048	0.00048828125	68
12	4,096	0.000244140625	74
13	8,192	0.0001220703125	80
14	16,384	0.00006103515625	86
15	32,768	0.000030517578125	92
16	65,536	0.0000152587890625	98
17	131,072	0.00000762939453125	104
18	262,144	0.000003814697265625	110
19	524,288	0.0000019073486328125	116
20	1,048,576	0.00000095367431640625	122

converter will beneficially minimize both quantization noise and A/D device error and provide increased converter linearity.

$$\text{Quantization error, Eqe} = \left(\frac{1}{q} \int_{-q/2}^{q/2} \varepsilon^2 \bullet d\varepsilon \right)^{1/2} = \frac{q}{2\sqrt{3}} \text{ rms volts} \qquad (5.6)$$

$$\text{Quantization quality, SNR} = 10 \log \left(\frac{V_{FS}/2 \sqrt{2}}{\text{Eqe}} \right)^2$$

$$= 10 \log \left(\frac{2^n \bullet q/2 \sqrt{2}}{q/2 \sqrt{3}} \right)^2 \qquad (5.7)$$

$$= 6.02 \, n + 1.76 \text{ power dB}$$

$$\text{Dynamic range} = 20 \log (2^n) = 6.02 \, n \text{ voltage dB} \qquad (5.8)$$

The input comparator is critical to the conversion speed and accuracy of an A/D converter, as shown in Figure 5.14. Generally, it must possess sufficient gain and bandwidth to achieve switching and settling to the amplitude error of interest, ultimately determined by noise sources present, such as described in Section 4.1.

Described now are seven prevalent A/D conversion methods and their application considerations. Architectures presented include integrating dual slope, sampling successive approximation, digital angle converters, charge

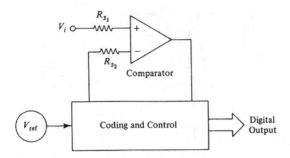

Figure 5.14 Comparator-oriented A/D converter diagram.

balancing and its evolution to oversampling sigma-delta converters, simultaneous or flash, and pipelined subranging. The performance of these conversion methods benefits from circuit advances and monolithic technologies in their accuracy, stability, and reliability that permit expression in terms of simplified static, dynamic, and temperature parameter error budgets, as illustrated by Table 5.8.

Quantizing uncertainty constitutes converter dynamic amplitude error, illustrated by Figure 5.12(b). Mean integral nonlinearity describes the maximum deviation of the static-transfer characteristic between initial and final code transitions in Figure 5.12(a). Circuit offset, gain, and linearity temperature coefficients are combined into a single percent of full-scale temperature coefficient expression. Converter signal-to-noise plus distortion expresses the quality of spurious and linearity dynamic performance. This latter error is influenced by data converter –3-dB frequency response, which generally must equal or exceed its conversion rate f_s to avoid amplitude and phase errors, considering the presence of input signal bandwidth values up to the $f_s/2$ folding frequency and the provisions of Tables 3.5 and 3.6. It is notable from Table 5.8 that the sum of the mean and RSS of converter errors provides a digital accuracy whose effective number of bits is typically 1 bit less than the specified converter wordlength.

Integrating converters provide noise rejection for the input signal at an attenuation rate of –20 dB per decade of frequency, as described in Figure 5.15, with sinc nulls at multiples of the integration period T. The ability of an integrator to provide this response is evident from its frequency response, $H(\omega)$, obtained by the integration of its impulse response, $h(t)$, in equation (5.9). Note that this noise improvement requires integration of the signal plus noise during

Table 5.8 Representative 12-Bit A/D Converter Errors

Mean integral nonlinearity (1 LSB)		0.024%
Quantizing uncertainty ($\frac{1}{2}$ LSB)		0.012
Temperature coefficient (1 LSB)		0.024
Noise + distortion		0.001
$\varepsilon_{A/D}$	$\Sigma mean + 1\sigma RSS$	0.050%

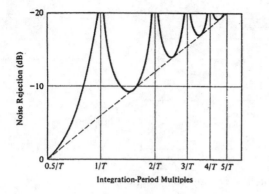
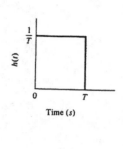

Figure 5.15 Integrating converter noise rejection.

the conversion period and consequently is not furnished when an S/H device precedes the converter. A conversion period of $16\frac{2}{3}$ msec will provide a useful null to 60 Hz interference, for example.

$$H(\omega) = \int_0^T h(t) \bullet e^{-j\omega t} = e^{-j\omega T/2} \bullet \frac{\sin \omega T/2}{\omega T/2} \tag{5.9}$$

Integrating dual-slope converters perform A/D conversion by the indirect method of converting an input signal to a representative pulse sequence that is totaled by a counter. Features of this conversion technique include self-calibration to component temperature drift, use of inexpensive components in its mechanization, and multiphasic integrations yielding improved resolution of the zero endpoint, shown in Figure 5.16. Operation occurs in three steps. First, the autozero phase stores converter analog offsets on the integrator with the input grounded. In the second phase an input signal is integrated for a fixed time, T_1. Finally the input is connected to a reference of opposite polarity and integration proceeds to zero during a variable time, T_2, within which clock pulses are totaled in proportion to the input signal amplitude. These operations are described by equations (5.10) and (5.11). Integrating converters are an early method the merits of which are best applied to narrow bandwidth signals such as encountered with hand-held multimeters. Wordlengths to 16 bits are available, but conversion is limited to 1 KSPS.

$$\Delta V_1 = \frac{1}{RC} \bullet V_i \bullet T_{1_{constant}} = \frac{1}{RC} \bullet V_{ref} \bullet T_{2\,variable} \tag{5.10}$$

$$T_2 = \frac{V_1 \bullet T_1}{V_{ref}} \tag{5.11}$$

The successive approximation technique is the most widely applied A/D converter type for computer interfacing primarily because its constant

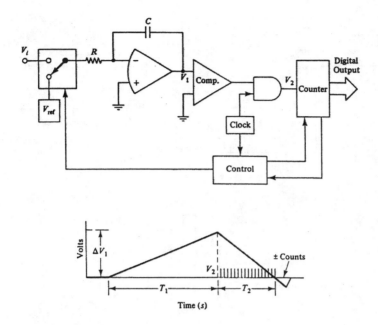

Figure 5.16 Dual-slope conversion.

conversion period, T, is independent of input signal amplitude. However, it requires a preceding S/H to satisfy its requirement for a constant input signal. This feedback converter operates by comparing the output of an internal DAC with the input signal at a comparator, where each bit of the wordlength is sequentially tested during n equal time subperiods in the development of an output code representative of input signal amplitude. Converter linearity is determined by the performance of its internal D/A. Figure 5.17 describes the operation of a sampling successive approximation converter. The conversion period and S/H acquisition time combined determine the maximum conversion rate, as described in Figure 5.6. Successive approximation converters are well suited for converting arbitrary signals, including those that are nonperiodic, in multiplexed systems. Wordlengths of 16 bits are available at conversion rates to 1 MSPS.

A common method for representing angles in digital form is in natural binary weighting, where the MSB represents 180 degrees and the MSB-1 represents 90 degrees, as tabulated in Table 5.9. Digital synchro conversion, shown in Figure 5.18, employs a Scott-T transformer connection and ac reference to develop the signals defined by equations (5.12) and (5.13). Sine ϕ and cosine ϕ quadrature multiplications are achieved by multiplying DACs whose difference is expressed by equation (5.14). A phase-detected dc error signal, described by equation (5.15), then pulses an up/down counter to achieve a digital output corresponding to the synchro angle θ. Related

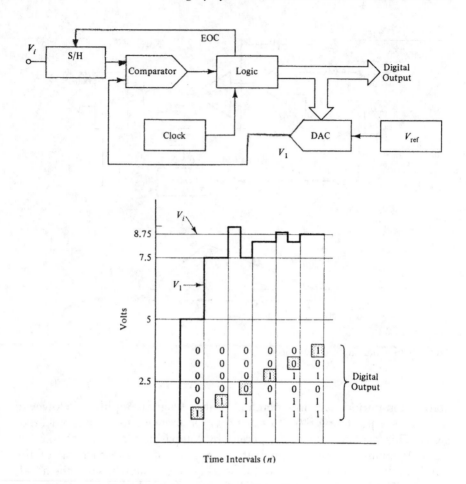

Figure 5.17 Successive approximation conversion.

Table 5.9 Binary Angle Representation

Bit	Degrees
1	180
2	90
3	45
4	22.5
5	11.25
6	5.625
7	2.812
8	1.406
9	0.703
10	0.351
11	0.176
12	0.088

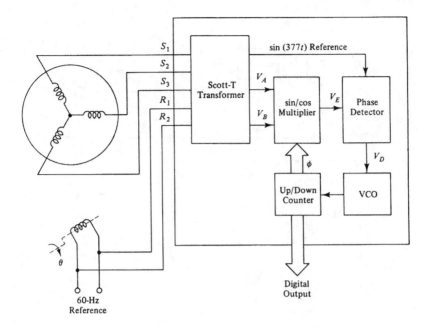

Figure 5.18 Synchro-to-digital conversion.

devices include digital vector generators, which generate quadrature circular functions as analog outputs from digital angular inputs.

$$V_A = \sin(377t) \bullet \sin \theta \qquad (5.12)$$

$$V_B = \sin(377t) \bullet \cos \theta \qquad (5.13)$$

$$V_E = \sin(377t) \bullet \sin(\theta - \phi) \qquad (5.14)$$

$$V_D = \sin(\theta - \phi) \qquad (5.15)$$

Charge-balancing A/D converters utilize a voltage-to-frequency circuit to convert an input signal to a current I_i from which is subtracted a reference current I_{ref}. This difference current is then integrated for successive intervals, with polarity reversals determined in one direction by a threshold comparator and in the other by clock count. The conversion period for this converter is constant, but the number of count intervals per conversion varies in direct proportion to input signal amplitude, as illustrated in Figure 5.19. Although the charge-balancing converter is similar in performance to the dual-slope converter, their applications diverge; the former is compatible with and integrated in microcontroller devices.

Sigma-delta conversion employs a version of the charge-balancing converter as its first stage to perform 1-bit quantization at an oversampled conversion rate fs whose "ones" density corresponds to analog input signal amplitude. The high quantizing noise resulting from 1-bit conversion is

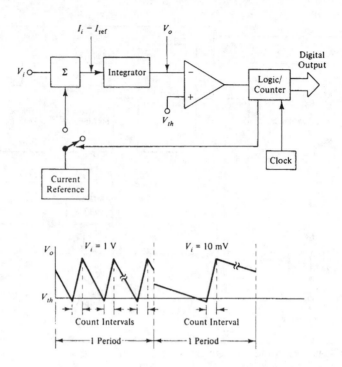

Figure 5.19 Charge-balancing conversion.

effectively spread over a wide bandwidth from the oversampling operation, which is amenable to efficient digital filtering because it is in the digital domain. The resulting spectrum is then resampled to an equivalent Nyquist-sampled signal bandwidth of n-bit resolution, shown in Figure 5.20. Sigma-delta converters are prevalent in medium-bandwidth, high-resolution periodic signal applications from measurement instruments to telecommunications and consumer electronics. Wordlengths to 20 bits for 100-kHz signal bandwidth are available. Because of signal latency associated with oversampling and decimation operations, however, sigma-delta converters are not compatible with multiplexed applications.

Simultaneous or flash converters are represented by the diagram of Figure 5.21, which require $2^n - 1$ comparators biased 1 LSB apart to encode an analog input signal to n-bit resolution. All quantization levels are simultaneously compared in a single clock cycle that produces a comparator "thermometer" code with a 1/0 boundary proportional to input signal amplitude. Comparator coding logic then provides a final digital output word. This architecture offers the fastest conversion rate achievable in a single clock cycle, but resolution is practically limited by the increasing number of comparators required for extending output wordlength. Wordlengths to 10 bits with 1023 comparators are available, however, at real time rates to 100 MSPS. The flash converter beneficially can accommodate

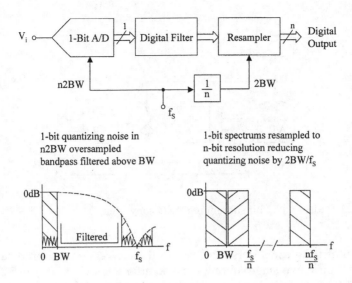

Figure 5.20 Sigma-delta conversion.

dynamic nonperiodic signals, like the slower successive approximation converter, but without an input S/H device. Applications include radar processors, electro-optical systems, and professional video.

Wideband and wide-range conversion are the province of pipelined subranging converters, which offer higher resolution than flash converters with nearly the same conversion rates. Wordlengths of 12 bits are common at conversion rates to 80 MSPS for applications ranging from digital spectrum analyzers to medical imaging. This architecture overcomes the comparator

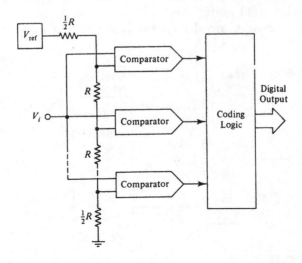

Figure 5.21 Simultaneous conversion.

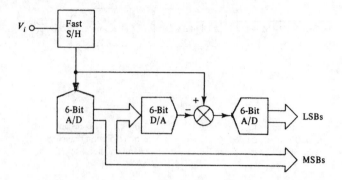

Figure 5.22 Wideband 12-bit subranging A/D converter.

limitation of flash converters by dividing the conversion task into cascaded stages. A typical two-stage subranging converter is shown in Figure 5.22, with two 6-bit flash A/D converters requiring only 126 comparators to provide a 12-bit wordlength, where the differential subrange is converted to LSB values by the second A/D. Flash converters of m bits in p stages offer a resolution of $p \times m$ bits with $p \times (2^m - 1)$ comparators. The throughput latency of p cycles of the pipeline impedes the conversion of nonperiodic signals, however.

Interrupt-initiated interfacing provides the flexibility required to accommodate asynchronous inputs from A/D converters. Upon the request of a peripheral controller, a processor interrupt is generated which initiates a service routine containing the device handler. This structure is illustrated by Figure 5.23 and offers enhanced throughput in comparison with program-initiated interfacing, previously described for DACs, by reconciling speed differences between processor buses and their peripheral controllers with FIFO buffers and DMA data transfers. Vectored multi-level-priority interrupts are also readily accommodated by this structure.

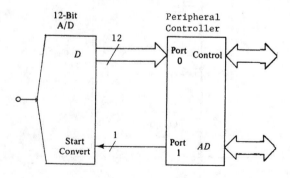

Figure 5.23 Interrupt-initiated A/D interface.

Bibliography

1. Sheingold, D.H., Ed., *Analog-Digital Conversion Handbook*, Analog Devices, Norwood, MA, 1972.
2. Artwick, B.A., *Microcomputer Interfacing*, Prentice Hall, Englewood Cliffs, NJ, 1980.
3. Bowers, D. "Analog Multiplexers: Their Technology and Operation," *Electronic Engineering*, Mid-September 1978, p. 23.
4. Bernstein, N., "What to Look for in Analog Input/Output Boards," *Electronics*, January 19, 1978, p. 113.
5. Gordon, B.M., *The Analogic Data-Conversion Systems Digest*, 2nd edition, Analogic, Wakefield, MA, 1977.
6. Hnatek, E.R., *A User's Handbook of D/A and A/D Converters*, John Wiley, New York, 1976.
7. Hoeschele, D.F., *Analog-to-Digital, Digital-to-Analog, Conversion Techniques*, John Wiley, New York, 1968.
8. Kester, W. "Test Video A/D Converters under Dynamic Conditions," *EDN*, August 18, 1982, p. 103.
9. Laplante, P.A., *Real-Time Systems Design and Analysis*, IEEE Press, New York, 1993.
10. Stone, H.S., *Microcomputer Interfacing*, Addison-Wesley, Reading, PA, 1982.
11. Bayes, G.S., ed., *Synchro and Resolver Conversion*, Analog Devices, Norwood, MA, 1980.
12. Zuch, E., *Data Acquisition and Conversion Handbook*, Datel-Intersil, Mansfield, MA, 1977.

chapter six

Sampled data
and recovery with
intersample error

6.0 Introduction

A fundamental requirement of sampled data systems is the sampling of continuous-time signals to obtain a representative set of numbers that can be utilized by a digital computer. The goal of this chapter is to provide an analytical understanding of this process. The first section explores theoretical aspects of sampling and the formal considerations of signal recovery, including ideal Wiener filtering in signal interpolation. Aliasing of signal and noise are considered next in a detailed development involving a heterodyne basis of evaluation. That development coordinates signal bandwidth, sample rate, and bandlimiting prior to sampling to achieve minimum aliasing error under conditions of significant aliasable content. The third section addresses intersample error in sampled systems and provides a sample-rate-to-signal-bandwidth ratio (f_s/BW), expressing the step-interpolator representation of sampled data in terms of equivalent binary accuracy. The fourth section derives a mean squared error criterion for evaluating the performance of practical signal recovery interpolator functions. This provides interpolated output signal quality in terms of the corresponding minimum required sample rate. The final section describes a video sampling and reconstruction example combining the foregoing sampled data considerations.

6.1 Sampled data theory

Observation of typical sensor signals generally reveals bandlimited continuous functions with diminished amplitude outside of a specific frequency band, except for interference or noise that may extend over a wide bandwidth. This is attributable to the natural rolloff or inertia associated with actual processes or systems providing the sensor excitation. Of interest is how much information is lost by the sampling operation and to what accuracy an original continuous signal can be reconstructed from its sampled values. The consideration of periodic sampling offers a mathematical solution to this problem for bandlimited sampled signals of bandwidth, BW. Signal discretization is illustrated for nonreturn-to-zero (NRZ) sampling and return-to-zero (RZ) sampling in Figure 6.1. This figure represents the two sampling classifications in both the time and frequency domains, where τ is the sampling function width and T the sampling period (the latter the inverse of sample rate f_s). The determination of specific sample rates that provide sampled data accuracies of interest is a central theme of this chapter.

The provisions of periodic sampling are based on Fourier analysis and include the existence of a minimum sample rate for which theoretically exact signal reconstruction is possible from the sampled sequence. This is significant in that signal sampling and recovery are considered simultaneously, correctly implying that the design of data conversion and recovery systems should also be considered jointly. The interpolation formula of equation (6.1) analytically describes the approximation, $\hat{x}(t)$, of a continuous-time signal, $x(t)$, with a finite number of samples from the sequence, $x(nT)$. $\hat{x}(t)$ is obtained from the inverse Fourier transform of the input sequence, which is derived from $x(t) \bullet p(t)$ as convolved with the ideal interpolation function, $H(f)$, of Figure 6.2. This results in the sinc amplitude response in the time domain owing to the rectangular characteristic of $H(f)$. Due to the orthogonal behavior of equation (6.1), only one nonzero term is provided at each sampling instant. Contributions of samples other than ones in the immediate neighborhood of a specific sample diminish rapidly because the amplitude response of $H(f)$ tends to decrease inversely with the value of n. Consequently, the interpolation formula provides a useful relationship for describing recovered bandlimited sampled data signals, with T chosen sufficiently small to prevent signal aliasing. Aliasing is discussed in detail in the following section. Figure 6.3 shows the behavior of this interpolation formula, including its output approximation, $\hat{x}(t)$

$$\hat{x}(t) = F^{-1}\{f[x(nT)] * H(f)\} \qquad (6.1)$$

$$= \sum_{n=-x}^{x} \left\{ T \int_{-BW}^{BW} x(nT)\, e^{-j2\pi f_n T} \right\} \bullet e^{j2\pi ft} \bullet df$$

$$= T \sum_{n=-x}^{x} x(nT) \frac{e^{j2\pi BW(t-nT)} - e^{-j2\pi BW(t-nT)}}{j2\pi BW(t-nT)}$$

$$= 2\,T\,BW \sum_{n=-x}^{x} x(nT) \frac{\sin 2\pi\, BW(t-nT)}{2\pi\, BW(t-nT)}$$

A formal description of this process was provided both by Wiener and Kolmogoroff. It is important to note that the ideal interpolation function, $H(f)$, utilizes both phase and amplitude information in reconstructing the

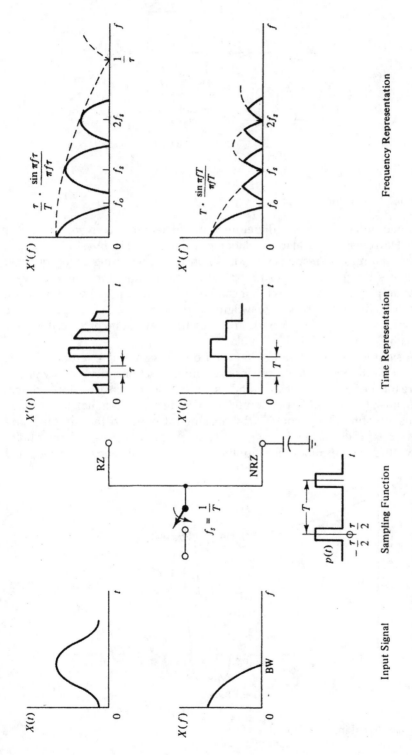

Figure 6.1 Sampled data time and frequency domain representation.

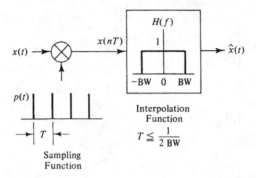

Figure 6.2 Ideal sampling and recovery.

recovered signal, $\hat{x}(t)$, and is therefore more efficient than conventional linear filters. However, this ideal interpolation function cannot be physically realized because its impulse response, $H(f)$, is noncausal, requiring an output that anticipates its input. As a result, practical interpolators for signal recovery utilize amplitude information that can be made efficient, although not optimum, by achieving appropriate weighting of the reconstructed signal. These principles are observed in Section 6.4 in the development of criteria for evaluating practical signal interpolators.

A significant consideration imposed on the sampling operation results from the finite width, τ, of practical sampling functions, denoted by $p(t)$ in Figure 6.1. Because the spectrum of a sampled signal consists of its original baseband spectrum, $X(f)$, plus a number of images of this signal, these image signals are shifted in frequency by an amount equal to the sampling frequency, f_s, and its harmonics, mf_s, as a consequence of the periodicity of $p(t)$. The width of τ determines the amplitude of these signal images, as attenuated

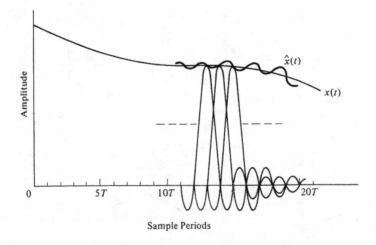

Figure 6.3 Signal interpolation.

by the sinc functions described by the dashed lines of $X(f)$ in Figure 6.1, for both RZ and NRZ sampling. Of particular interest is the attenuation impressed upon the baseband spectrum of $X(f)$ corresponding to the amplitude and phase of the original signal, $X(f)$. A useful criterion is to consider the average baseband amplitude error between dc and the signal BW, expressed as a percentage of the full-scale departure from unity gain. Also, process bandwidth must be sufficient to support these image spectra until their amplitudes are attenuated by the sinc function to preserve signal fidelity. The mean sinc amplitude error is expressed for RZ and NRZ sampling by equations (6.2) and (6.3). The sampled data bandwidth requirement for NRZ sampling is generally more efficient in system bandwidth utilization than the $1/\tau$ null provided by RZ sampling. The minimization of mean sinc amplitude error may also influence the choice of f_s. The folding frequency f_o in Figure 6.1 is an identity equal to $f_s/2$, and the specific NRZ sinc attenuation at f_o is always 0.636, or –3.93 dB.

$$\overline{\varepsilon_{\text{RZ sinc \%FS}}} = \frac{1}{2}\left(1 - \frac{\tau}{T} \cdot \frac{\sin \pi \, \text{BW}\tau}{\pi \, \text{BW}\tau}\right) \cdot 100\% \qquad (6.2)$$

$$\overline{\varepsilon_{\text{NRZ sinc \%FS}}} = \frac{1}{2}\left(1 - \frac{\sin \pi \, \text{BW}T}{\pi \, \text{BW}T}\right) \cdot 100\% \qquad (6.3)$$

RZ sampling is primarily used for multiplexing multichannel signals into a single channel, such as encountered in telemetry systems. Figure 6.1 provides that the dc component of RZ sampling has an amplitude of τ/T, its average value or sampling duty cycle, which may be scaled as required by the system gain. NRZ sampling is inherent in the operation of all data-conversion components encountered in computer input–output systems and reveals a dc component proportional to the sampling period T. In practice, this constant is normalized to unity by the $1/T$ impulse response associated with the transfer functions of actual data-conversion components.

Note that the sinc function and its attenuation with frequency in a sampled data system is essentially determined by the duration of the sampled signal representation $X(t)$ at any point of observation, as illustrated in Figure 6.1. For example, an A/D converter with a conversion period, T, double the value employed for a following connected D/A converter will exhibit an NRZ sinc function with twice the attenuation rate vs. frequency as that of the D/A, which is attributable to the transformation of the conversion period. Such D/A oversampling accordingly offers reduced output sinc error, discussed in Section 6.4. Note that sampled data systems, therefore, possess a single sinc function that is transformed as a function of changes in sampling parameters.

6.2 Aliasing of signal and noise

The effect of undersampling a continuous signal is illustrated in both the time and frequency domains in Figure 6.4. This demonstrates that the mapping of a signal to its sampled data representation does not have an identical reverse

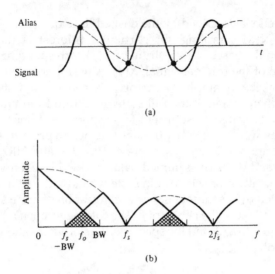

Figure 6.4 Time and frequency representation of undersampled signal aliasing.

mapping if it is reconstructed as a continuous signal when it is undersampled. Such signals appear as lower frequency aliases of the original signal and are defined by equation (6.4) when $f_s < 2$ BW. As the sample rate, f_s, is reduced, samples move further apart in the time domain and signal images closer together in the frequency domain. When image spectrums overlap as illustrated in Figure 6.4(b), signal aliasing occurs. The consequence is the generation of intermodulation distortion that cannot be removed by later signal processing operations. Of interest is aliasing at f_o between the baseband spectrum—epresenting the amplitude and phase of the original signal—and the first image spectrum. The folding frequency, f_o, is the highest frequency at which sampled data signals may exist without being undersampled. Accordingly, f_s must be chosen greater than twice the signal BW to ensure the absence of signal aliasing, which usually is readily achieved in practice.

$$f_{alias} = [f_s - BW] \qquad f_s < 2 \text{ BW}$$
$$= \text{nonexistent} \quad f_s \geq 2 \text{ BW} \qquad (6.4)$$

Of greater general concern and complexity is noise aliasing in sampled data systems. This involves either out-of-band signal components, such as coherent interference, or random noise spectra, present above f_o and therefore undersampled. One or both of these sources are frequently present in sampled data systems. Consequently, the design of these systems should provide for the analysis of noise aliasing and the coordination of system parameters to achieve the aliasing attenuation of interest. Understanding of baseband aliasing is aided with reference to Figures 6.5 and 6.6. The noise aliasing source bands shown are heterodyned within the baseband signal between dc and f_o, derived by equation (6.5) as $mf_s - BW \leq f_{noise} < mf_s + BW$, as a

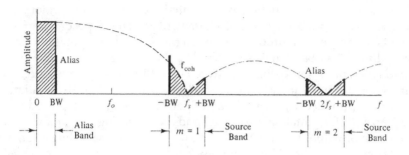

Figure 6.5 Coherent interference aliasing.

consequence of the sampling function spectra that arise at multiples of f_s. The resulting combination of signal and aliasing components generates intermodulation distortion proportional to the baseband alias amplitude error derived by equations (6.7) through (6.10).

Coherent alias frequencies capable of interfering with baseband signals are defined by equation (6.6). The amplitude of the aliasing error components expressed as a percent of full scale are provided for both NRZ and RZ sampling by equation (6.7) with the appropriate sinc function argument. Note that this equation may be evaluated to determine the aliasing amplitude error with or without presampling filtering and its effect on aliasing attenuation. For example, consider a 1-Hz signal bandwidth for an NRZ sampled data system with an f_s of 24 Hz. A 23-Hz coherent interfering input signal of –6 dB amplitude (50%FS) will be heterodyned both to 1 Hz and 47 Hz by this 24-Hz sampling frequency, with negligible sinc attenuation at 1 Hz and approximately –30 dB at 47 Hz, for a coherent baseband aliasing error of 50%FS, applying equation (6.7) in the absence of a presampling filter. This is illustrated by Figure 6.5. The addition of a lowpass three-pole Butterworth <u>presampling</u> filter with a 3-Hz cutoff frequency to minimize filter error to 0.1%FS over the signal bandwidth then provides –52 dB attenuation to the 23-Hz interfering signal for a negligible 0.12%FS baseband aliasing error, shown by the calculations accompanying equation (6.7). This filter may be visualized superimposed on that figure.

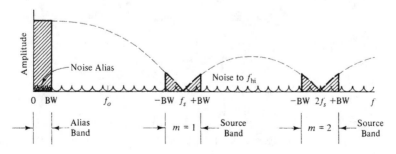

Figure 6.6 Random interference aliasing.

A more complex situation is presented in the case of random noise because of its wideband spectral characteristic. This type of interference exhibits a uniform amplitude representing a Gaussian probability distribution. Aliased baseband noise power, N_{alias}, is determined as the sum of heterodyned noise source bands between $mf_s - BW$ noise. These bands occur at intervals of f_s in frequency, shown in Figure 6.6 up to a –3-dB bandlimiting f_{hi}, such as provided by an input amplifier cutoff frequency preceding the sampler, with f_{hi}/f_s total noise source bands contributing. N_{alias} may be evaluated with or without the attenuation provided by a presampling filter, $A(f)$, in determining baseband random noise aliasing error, which is expressed as an aliasing signal-to-noise ratio (SNR) in equations (6.9) and (6.10). The small sinc amplitude attenuation at baseband is omitted for simplicity.

Consider a –20-dB example V_{noise} rms level extending from dc to an f_{hi} of 1 kHz. Solution of equations (6.8) through (6.10), in the absence of a filter with $A(f) = 1.0$, yields 0.42 V full-scale square (watts) into 1 Ω as N_{alias} with an f_s of 24 Hz and 42 source bands contributing to 1 kHz for a random noise aliasing error of 90%FS. Consideration of the previous 1-Hz signal bandwidth and 3-Hz cutoff three-pole Butterworth lowpass filter provides –54 dB average attenuation over the first noise source band centered at f_s. Significantly greater filter attenuation is imposed at higher noise frequencies, resulting in negligible contribution from those source bands to N_{alias}. The presampling filter effectiveness, therefore, is such that the random noise aliasing error is only 0.027%FS. It is notable that baseband aliased noise inversely follows the response of $A(f)$ to f_o, owing to heterodyned noise contributions from f_o to f_s.

$$mf_s - BW \quad f_{noise} < mf_s + BW \qquad \text{alias source frequencies} \qquad (6.5)$$

$$f_{coherent \atop alias} = |mf_s - f_{coh}| \qquad \text{at baseband} \qquad (6.6)$$

$$= 24 \text{ Hz} - 23 \text{ Hz}$$
$$= 1 \text{ Hz } (m = 1)$$

$$\varepsilon_{coherent \atop alias} = V_{coh\%FS} \bullet \text{filter attn} \bullet \text{sinc} \qquad (6.7)$$

$$= 50\%FS \bullet \frac{1}{\sqrt{1 + \left(\dfrac{f_{coh}}{f_c}\right)^{2n}}} \bullet \text{sinc}\left(\frac{|mf_s - f_{coh}|}{fs}\right)$$

$$= 50\%FS \bullet \frac{1}{\sqrt{1 + \left(\dfrac{23}{3}\right)^{6}}} \bullet \text{sinc}\left(\frac{|24 - 23|}{24}\right)$$

$$= 50\%FS \bullet (0.0024) \bullet (0.998)$$

$$= 0.12\%FS \qquad \text{with presampling filter}$$

$$N_{alias} = \sum_{0}^{\overset{\text{\#source}}{\text{bands}}} (V_{noise}\text{rms})^2 \bullet A^2(f) \qquad \text{at baseband} \qquad (6.8)$$

$$= \sum_{0}^{f_{hi}/f_s} (0.1\, V_{FS})^2 \bullet \left[\frac{1}{\sqrt{1+\left(\dfrac{f_s}{f_c}\right)^{2n}}} \right]^2$$

$$= \sum_{0}^{1} (0.01\, V_{FS}^2) \bullet \left[\frac{1}{\sqrt{1+\left(\dfrac{24}{3}\right)^{6}}} \right]^2$$

$$= 0.038 \times 10^{-6} \bullet V_{FS}^2 \text{ watt into } 1\,\Omega$$

$$\text{SNR}_{\underset{alias}{random}} = \frac{V_s^2\ \text{rms}}{N_{alias}} \qquad\qquad (6.9)$$

$$\varepsilon_{\underset{alias}{random}} = \frac{\sqrt{2}\bullet 100\%}{\sqrt{\text{SNR}_{\underset{alias}{random}}}} = \frac{\sqrt{2}\bullet 100\%}{\sqrt{V_{FS}^2/0.038 \times 10^{-6} V_{FS}^2}} \qquad (6.10)$$

$$= 0.027\%\text{FS} \qquad \text{with presampling filter}$$

Table 6.1 offers an efficient coordination of presampling filter specifications employing a conservative criterion of achieving –40-dB input attenuation at f_o in terms of a required f_s/BW ratio that defines the minimum sample rate for preventing noise aliasing. The foregoing coherent and random noise aliasing

Table 6.1 Coordination of Sample Rate, Signal Bandwidth, and Sinc Function with Presampling Filter for Aliasing Attenuation at the Folding Frequency

				f_s/BW for –40-dB attenuation at f_o including –4-dB sinc with filter f_c of			Filter $\overline{\varepsilon_{\%FS}}$ per signal type	
Application	RC	Bessel	Butterworth	20 BW	10 BW	3 BW	DC, Sines	Harmonic
DC signals	1			2560			0.10	1.20
Linear phase		3			80		0.10	0.10
General			3			24	0.10	0.11
Brickwall			6			12	0.05	0.15

examples meet these requirements, with their f_s/BW ratios of 24 employing the general-application three-pole Butterworth presampling filter, whose cut-off frequency, f_c, of three times signal bandwidth provides only a nominal device error addition while achieving significant antialiasing protection. RC presampling filters are clearly least efficient and appropriate only for dc signals considering their required f_s/BW ratio to obtain useful aliasing atten-uation and nominal error. Six-pole Butterworth presampling filters are most efficient in conserving sample rate while providing equal aliasing attenuation at the cost of greater filter complexity. A three-pole Bessel filter is unparalleled in its linearity to both amplitude and phase for all signal types as an anti-aliasing filter, but it requires an inefficient f_s/BW ratio to compensate for its passband amplitude rolloff. The following sections consider the effect of sample rate on sampled data accuracy—first as step-interpolated data prin-cipally encountered on a computer data bus and then including postfilter interpolation associated with output signal reconstruction.

6.3   Sampled data intersample and aperture errors

The NRZ-sampling step-interpolated data representation of Figure 6.7 denotes the way converted data are handled in digital computers, whereby the present sample is current data until a new sample is acquired. Both intersample and aperture volts, ΔV_{pp} and $\Delta V'_{pp}$, respectively, are derived in this development as time–amplitude relationships to augment that understanding.

In real-time data conversion systems, the sampling process is followed by quantization and encoding, all of which are embodied in the A/D conversion process described by Figure 5.11. Quantization is a measure of the number of discrete amplitude levels that may be assigned to represent a signal waveform, and is proportional to A/D converter output wordlength in bits. A/D quan-tization levels are uniformly spaced between 0 and V_{FS}, with each being equal to the LSB interval as described in Figure 5.12. For example, a 12-bit A/D converter provides a quantization interval proportional to 0.024%FS. This typical converter wordlength thus provides quantization that is sufficiently small to permit independent evaluation of intersample error without the influ-ence of quantization effects even though the source of both is the A/D con-verter. Note that both intersample and aperture error are system errors, whereas quantization error is a part of the A/D converter device error.

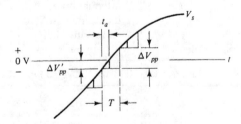

Figure 6.7 Intersample and aperture error representation.

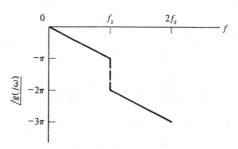

Figure 6.8 Step-interpolator phase.

NRZ sampling is inherent in the operation of sample-hold (S/H), analog-to-digital (A/D), and digital-to-analog (D/A) devices by virtue of their step-interpolator sampled data representation. Equation (6.11) describes the impulse response for this data representation in the derivation of a frequency-domain expression for step-interpolator amplitude and phase. Evaluation of the phase term at the sample rate, f_s, discloses that an NRZ-sampled signal exhibits an average time delay equal to $T/2$ with reference to its input. This linear phase characteristic is illustrated in Figure 6.8. The sampled input signal is acquired as shown in Figure 6.9(a) and represented as discrete values in a digitally encoded form. Figure 6.9(b) describes the average signal delay with reference to its input of Figure 6.9(a). The difference between this average signal and its step-interpolator representation constitute the peak-to-peak intersample error constructed in Figure 6.9(c).

$$g(t) = U(t) - U(t - T) \qquad (6.11)$$

$$g(s) = \frac{1}{s} - \frac{e^{-T}}{s}$$

$$g(j\omega) = \frac{1 - e^{-j\omega T}}{j\omega}$$

$$= T\frac{\sin \pi f T}{\pi f T}\underline{/-j\omega T/2} \quad \text{NRZ impulse response}$$

Evaluating delay at $f_s = \dfrac{1}{T}$:

$$\underline{/g(j\omega)} = -\pi = 2\pi f t$$

$$\therefore t = -\frac{T}{2}\ \text{sec} \qquad \text{sampled signal delay}$$

Equation (6.12) describes the intersample volts, ΔV_{pp}, for an instantaneous sinusoidal signal V_s evaluated at its maximum rate-of-change zero crossing, shown in Figure 6.7. This representation is converted to an equivalent ΔV_{rms} through normalization by $2\sqrt{5}$ from multiplying the $2\sqrt{2}$ sinusoidal pp-rms factor and the $\sqrt{2.5}$ crest factor triangular step-interpolation contribution of Figure 6.9(c). This expression is also equal to the square root of mean squared

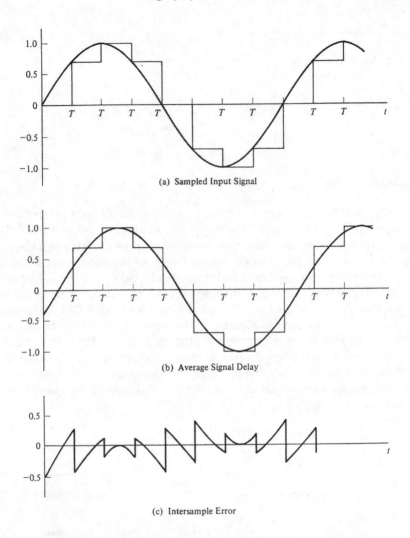

(a) Sampled Input Signal

(b) Average Signal Delay

(c) Intersample Error

Figure 6.9 Step-interpolator signal representation.

error (MSE), which diminishes as the true signal and its sampled data representation converge. Equation (6.13) re-expresses equation (6.12) to define a more useful amplitude error, $\varepsilon_{\Delta V\%FS}$, represented in terms of binary equivalent values in Table 6.2, and is then rearranged in terms of a convenient f_s/BW ratio for application purposes. Describing the signal V_s relative to the specific V_{FS} scaling also permits accommodation of the influence of signal amplitude on the representative rms intersample error of a digitized waveform. Intersample error thus represents the departure of A/D output data from their corresponding continuous input signal values as a consequence of converter sampling, quantizing, and encoding functions and signal bandwidth and amplitude dynamics. For example, a signal V_S of one-half V_{FS} provides only half the intersample error obtained at full V_{FS}, for a constant signal bandwidth and sample rate,

Table 6.2 Step-Interpolated Sampled Data Equivalents

Binary Bits (accuracy)	Intersample Error $\varepsilon_{\Delta V\%FS}$ (1 LSB)	f_s/BW (numerical)	Applications
0	100.0	2	Nyquist limit
1	50.0		
2	25.0		
3	12.5		
4	6.25	32	Digital toys
5	3.12		
6	1.56		
7	0.78		
8	0.39	512	Video systems
9	0.19		
10	0.097		
11	0.049		
12	0.024	8192	Industrial I/O
13	0.012		
14	0.006		
15	0.003		
16	0.0015	131,072	Instrumentation
17	0.0008		
18	0.0004		
19	0.0002		
20	0.0001	2,097,152	High-end audio

$$\Delta V_{pp} = T \bullet \frac{dV_s}{dt} \text{ intersample volts} \tag{6.12}$$

$$= T \bullet \frac{d}{dt} V_s \sin 2\pi \, BWt \Big|_{t=0} = 2\pi \, T \, BW \, V_{s_{pk}}$$

$$\Delta V_{rms} = \frac{2\pi \, T \, BW \, V_{s_{pk}}}{2\sqrt{5}} = \sqrt{MSE} \text{ volts}$$

$$\varepsilon_{\Delta V\%FS} = \frac{\Delta V_{rms}}{V_{FS_{pk}}/\sqrt{2}} \bullet 100\% \text{ intersample error} \tag{6.13}$$

$$= \frac{\sqrt{2} \, \pi \, BW \, V_{s_{pk}}}{\sqrt{5} \, f_s \, V_{FS_{pk}}} \bullet 100\%$$

$$\frac{f_s}{BW} = \frac{\sqrt{2} \, \pi \, V_{s_{pk}} \, 100\%}{\sqrt{5} \, \varepsilon_{\Delta V\%FS} \, V_{FS_{pk}}} \text{ step-interpolated data}$$

because an equal number of encoded amplitude values are distributed over only half the signal excursion.

Determining the step-interpolated intersample error of interest is aided by Table 6.2 and equation (6.13). For example, 8-bit binary accuracy requires an f_s/BW ratio of 512 considering its LSB amplitude value of

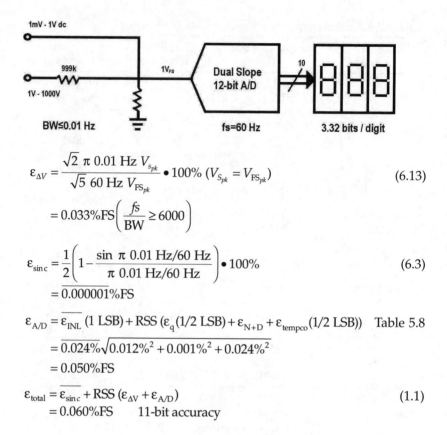

$$\varepsilon_{\Delta V} = \frac{\sqrt{2} \; \pi \; 0.01 \; \text{Hz} \; V_{S_{pk}}}{\sqrt{5} \; 60 \; \text{Hz} \; V_{FS_{pk}}} \bullet 100\% \; (V_{S_{pk}} = V_{FS_{pk}}) \tag{6.13}$$

$$= 0.033\% \text{FS} \left(\frac{fs}{\text{BW}} \geq 6000 \right)$$

$$\varepsilon_{\sin c} = \frac{1}{2} \left(1 - \frac{\sin \pi \; 0.01 \; \text{Hz}/60 \; \text{Hz}}{\pi \; 0.01 \; \text{Hz}/60 \; \text{Hz}} \right) \bullet 100\% \tag{6.3}$$

$$= \overline{0.000001}\% \text{FS}$$

$$\varepsilon_{A/D} = \overline{\varepsilon_{\text{INL}} \; (1 \; \text{LSB}) + \text{RSS} \; (\varepsilon_q (1/2 \; \text{LSB}) + \varepsilon_{N+D} + \varepsilon_{\text{tempco}} (1/2 \; \text{LSB}))} \quad \text{Table 5.8}$$

$$= \overline{0.024\% \sqrt{0.012\%^2 + 0.001\%^2 + 0.024\%^2}}$$

$$= 0.050\% \text{FS}$$

$$\varepsilon_{\text{total}} = \overline{\varepsilon_{\sin c} + \text{RSS} \; (\varepsilon_{\Delta V} + \varepsilon_{A/D})} \tag{1.1}$$

$$= 0.060\% \text{FS} \qquad \text{11-bit accuracy}$$

Figure 6.10 Digital dc voltmeter.

0.39%FS. This implies sampling a sinusoid uniformly every 0.77 degree with the waveform peak amplitude scaled to the full-scale value. This obviously has an influence on the design of sampled data systems and the allocation of their resources to achieve an intersample error of interest. With harmonic signals the tenth-harmonic amplitude value typically declines to one-tenth that of the fundamental frequency amplitude such that intersample error remains constant between these signal frequencies for arbitrary sample rates. The f_s/BW ratio of two provides an intersample error reference defining frequency sampling that is capable of quantifying only signal polarity changes. Unlike digital measurement and control systems, where quantitative amplitude accuracy is of interest, frequency sampling is employed for information that is encoded in terms of signal frequencies as encountered in communications systems that usually involve qualitative interpretation. For example, digital telephone systems often employ 7-bit accuracy, to meet a human sensory error/distortion perception threshold generally taken as 0.7%FS.

Figure 6.10 describes an elementary digital error budget example of 11-bit binary accuracy that is ample for a 3-decimal-digit dc digital voltmeter whose step-interpolated 3.32 bits/digit requires only 10 bits for display. This acquisition system can accommodate a signal bandwidth to 10 mHz at a sample rate of 60 Hz with an f_s/BW of 6000. From Chapter 5, intrinsic noise rejection of the integrating A/D converter beneficially provides amplitude nulls to possible voltmeter interference at the f_s value of 60 Hz and –20 dB per decade rolloff to other input frequencies.

Aperture time, t_a, describes the finite amplitude uncertainty $\Delta V'_{pp}$ within which a sampled signal is acquired such as by a sample-hold device, referencing Figure 6.7 and equation (6.14), that involves the same relationships expressed in equation (6.12). Otherwise, sampling must be accomplished by a device whose performance is not affected by input signal change during acquisition, such as an integrating A/D. In that direct conversion case, t_a is identically the sampling period T. A principal consequence of aperture time is the superposition of an additional sinc function on the sampled data spectrum. The mean aperture error over the baseband signal described by equation (6.15), however, is independent of the mean sinc error defined by equation (6.3). Although intersample and aperture performance are similar in their relationships, variation in t_a has no influence on intersample error. For example, a fast sample-hold preceding an A/D converter can provide a small aperture uncertainty, but intersample error continues to be determined by the sampling period T. Figure 6.11 is a nomograph of equation (6.15) that describes aperture error in terms of binary accuracy. Aperture error is negligible in contemporary data conversion systems and consequently not included in the error summary.

$$\Delta V'_{pp} = 2\pi t_a \ \mathrm{BW} \ V_{s_{pk}} \ \text{aperture volts} \tag{6.14}$$

$$\overline{\varepsilon_{a\%FS}} = \frac{1}{2}\left(1 - \frac{\sin \pi \ \mathrm{BW} \ t_a}{\pi \ \mathrm{BW} \ t_a}\right) \bullet 100\% \tag{6.15}$$

6.4   Output signal interpolation functions

The recovery of continuous analog signals from discrete digital signals is required in the majority of instrumentation applications. Signal reconstruction may be viewed from either time domain or frequency domain perspectives. In time domain terms, recovery is similar to interpolation techniques in numerical analysis involving the generation of a locus that reconstructs a signal by connecting discrete data samples. In the frequency domain, efficient signal recovery involves bandlimiting a D/A output with a lowpass postfilter to attenuate image spectra present above the baseband signal. It is of further interest to pursue signal reconstruction methods that are more efficient in sample rate requirements than the step-interpolator signal representation described by Table 6.2.

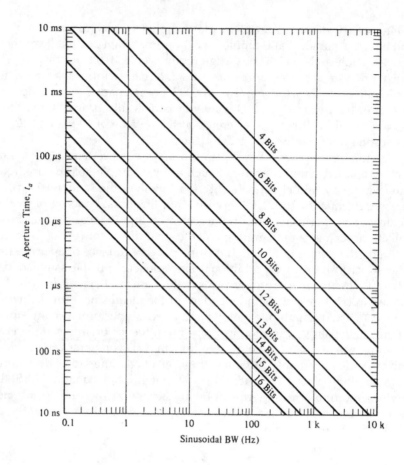

Figure 6.11 Aperture binary accuracy nomograph.

Figure 6.12 illustrates direct-D/A signal recovery with extensions that add both linear interpolator and postfilter functions. Signal delay is problematic in digital control systems such that a direct-D/A output is employed with image spectra attenuation achieved by the associated process closed-loop bandwidth. This example is evaluated in Chapter 9. Linear interpolation is a capable reconstruction function, but achieving a nominal interpolator device error is problematic. Linear interpolator effectiveness is defined by first-order polynomials whose line-segment slopes describe the difference between consecutive data samples.

Figure 6.13 shows a frequency domain representation of a sampled signal of bandwidth BW with images about the sampling frequency, f_s. This ensemble illustrates image spectra attenuated by the sinc function and lowpass postfilter in achieving convergence of the total sampled data spectra to its ideal baseband BW value. An infinite-series expression of the image spectra summation is given by equation (6.16) that equals the MSE for direct D/A output.

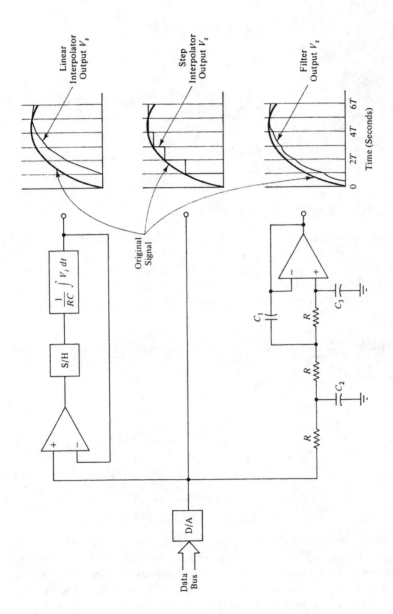

Figure 6.12 Signal reconstruction methods.

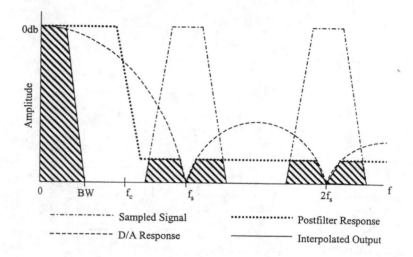

- - - - - - - Sampled Signal   •••••••••••••••• Postfilter Response
- - - - - - - - - D/A Response   ———————— Interpolated Output

Figure 6.13 Signal recovery spectral ensemble.

It follows that step-interpolated signal intersample error may be evaluated by equation (6.17), employing this MSE in deriving the first output interpolator function of Table 6.3, whose result corresponds identically to that of equations (6.12) and (6.13). Note that the sinc terms of equation (6.17) are evaluated at the worst-case first image maximum amplitude frequencies of $f_s \pm BW$.

$$\text{MSE} = V_s^2 \sum_{k=1}^{\infty} \left[\text{sinc}^2\left(k - \frac{BW}{f_s}\right) + \text{sinc}^2\left(k + \frac{BW}{f_s}\right) \right] \text{D/A output} \qquad (6.16)$$

$$= 2V_s^2 \left[\text{sinc}^2\left(1 - \frac{BW}{f_s}\right) + \text{sinc}^2\left(1 + \frac{BW}{f_s}\right) \right]$$

$$\varepsilon_{\Delta V\%FS} = \left[\frac{V_{OFS}^2}{2V_s^2 \left[\text{sinc}^2\left(1 - \frac{BW}{f_s}\right) + \text{sinc}^2\left(1 + \frac{BW}{f_s}\right) \right]} \right]^{-1/2} \bullet 100\% \qquad (6.17)$$

The choice of interpolator function should include a comparison of realizable signal intersample error and the error addition provided by the interpolator device, with the goal of realizing not greater than parity in these values. Figure 6.14 shows a comparison of four output interpolators for a sampled sinusoidal signal at a modest f_s/BW ratio of 10. The three-pole Butterworth postfilter is especially versatile for image spectra attenuation with dc, sinusoidal, and harmonic signals and adds only nominal device error with reference to Tables 3.5 and 3.6. Its 6-bit improvement over direct-D/A recovery is substantial, with significant convergence toward ideal signal

Table 6.3 Output Interpolator Functions

Interpolator	Amplitude	Intersample Error $\varepsilon_{\Delta V\%FS}$
D/A	$\text{sinc}\,(f/f_s)$	$\left[\dfrac{2V_s^2\left[\text{sinc}^2\left(1-\dfrac{BW}{f_s}\right)+\text{sinc}^2\left(1+\dfrac{BW}{f_s}\right)\right]}{V_{O_{FS}}^2}\right]^{-1/2}\cdot 100\%$
D/A + linear	$\text{sinc}^2\,(f/f_s)$	$\left[\dfrac{V_s^2\left[\text{sinc}^4\left(1-\dfrac{BW}{f_s}\right)+\text{sinc}^4\left(1+\dfrac{BW}{f_s}\right)\right]}{V_{O_{FS}}^2}\right]^{-1/2}\cdot 100\%$
D/A + 1-pole RC	$\text{sinc}\,(f/f_s)\left[1+(f/f_c)^2\right]^{-1/2}$	$\left[V_s^2\left[\text{sinc}^2\left(1-\dfrac{BW}{f_s}\right)\right]\left[1+\left(\dfrac{f_s-BW}{f_c}\right)^{2n}\right]^{-1}\right.$ $\left.+\,\text{sinc}^2\left(1+\dfrac{BW}{f_s}\right)\left[1+\left(\dfrac{f_s+BW}{f_c}\right)^{2n}\right]^{-1}\right]^{-1/2}\cdot 100\%$
D/A + Butterworth n-pole lowpass	$\text{sinc}\,(f/f_s)\left[1+(f/f_c)^{2n}\right]^{-1/2}$	

$f_s \pm BW$ substituted for f

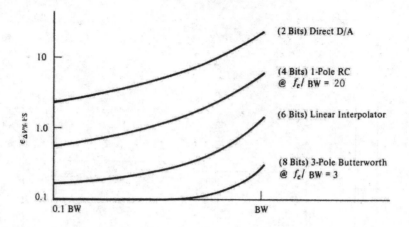

Figure 6.14 Output interpolator comparison (sinusoid, fs/BW = 10, Vs = V_{FS}).

reconstruction. Bessel filters require an f_c/BW of 20 to obtain a nominal device error, which limits their effectiveness to that of a one-pole RC. However, this does not diminish the utility of Bessel filters for reconstructing phase signals. Interpolator residual intersample error values revealed at fractional band-width in Figure 6.14 also define the equivalent data wordlengths necessary to preserve the interpolated signal accuracy achieved. Table 6.4 describes the signal time delay encountered in transit through the respective interpolation functions.

  Oversampled data conversion, introduced by the sigma-delta A/D con-verter of Figure 5.20, relies on the increased quantization SNR of 6 dB for each fourfold increase in f_s, enabling one binary-bit-equivalent of additional perfor-mance from Table 5.7. The merit of oversampled D/A conversion, compared to Nyquist sampling and recovery where signal BW may exist up to the folding frequency value of $f_s/2$, is a comparable output SNR improvement without increasing the converter wordlength accompanied by reduced sinc attenua-tion with reference to equation (6.3). Figure 6.15 shows the performance improvement of four times oversampling D/A conversion with the sampled signal present every fourth sample. First, the fixed quantization noise power

Table 6.4 Interpolation Transfer Delay

Interpolator	Time (sec)
D/A	$\dfrac{1}{2f_s}$
D/A + linear	$\dfrac{1}{2f_s} + \dfrac{1}{f_s}$
D/A + Butterworth n-pole	$\dfrac{1}{2f_s} + \dfrac{n}{4f_c}$

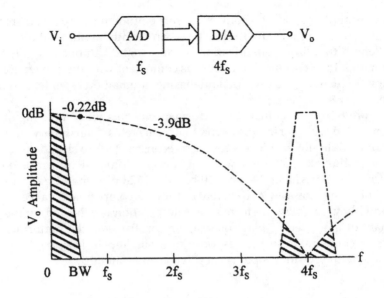

Figure 6.15 Oversampled D/A conversion spectrum.

for any D/A wordlength is now distributed over four times the spectral occupancy such that only one-fourth of this noise is in the signal bandpass up to bandwidth, BW yielding a 6-dB SNR improvement. Further, the accompanying sinc amplitude attenuation at signal bandwidth is –0.22 dB compared to –3.9 dB encountered with Nyquist sampled signals. Equivalent oversampling performance additions may be obtained at 16 f_s, 64 f_s, and higher multiples. However, although oversampling achieves reduced quantization noise and sinc attenuation, it cannot increase signal data content.

6.5   Video sampling and reconstruction

Industrial machine vision, laboratory spectral analysis, and medical imaging instrumentation are all supported by advances in digital signal processing that are frequently coupled to television standards and computer graphics technology. Real-time imaging systems usefully employ line-scanned television standards such as RS-343A and RS-170 that generate 30 frames per second, with 525 lines per frame interlaced into one even-line and one odd-line field per frame. Each line has a sweep rate of 53.3 μsec, plus 10.2 μsec for horizontal retrace. The bandwidth required to represent discrete picture elements (pixels) is therefore determined by the discrimination of active and inactive pixels of equal width in time along a scanning line.

The implementation of a high-speed data conversion system is largely a wideband analog design task. Baseline considerations include employing data converters possessing intrinsic speed with low spurious performance. The example ADS822 A/D converter by Burr-Brown is capable of a

40-megasample-per-second conversion rate employing a pipelined architecture for input signals up to 10-MHz bandwidth, with a 10-bit output wordlength that limits quantization noise to –60 dB. A one-pole RC input filter with a 15-MHz cutoff frequency is coincident with the conversion rate folding frequency, f_o, to provide antialiasing attenuation of wideband input noise.

Figure 6.16 reveals that the performance of this video imaging system is dominated by intersample error that achieves a 5-bit binary accuracy, providing 32 luminance levels for each reconstructed pixel. A detailed system error budget, therefore, will not reveal additional influence on this performance. An Analog Devices 10-bit ADV7128 pipelined D/A converter with a high-impedance video current output is a compatible data reconstructor providing glitchless performance. Interpolation is achieved by the time constant of the video display for image reconstruction, the performance of which is comparable to the response of a single-pole lowpass filter constrained by the 30-frames-per-second television standard.

Video Interpolation

$$\varepsilon_{\Delta V} = \left[\frac{V_{O_{FS}}^2}{V_S^2 \cdot \left\{ \sin c^2 \left(1 - \frac{BW_{pixel}}{f_s} \right) \cdot \left[1 + \left(\frac{f_s - BW_{pixel}}{f_{phosphor}} \right)^2 \right]^{-1} + \sin c^2 \left(1 + \frac{BW_{pixel}}{f_s} \right) \cdot \left[1 + \left(\frac{f_s + BW_{pixel}}{f_{phosphor}} \right)^2 \right]^{-1} \right\}} \right]^{-1/2} \cdot 100\%$$

$$= \left[\frac{1V^2}{1V^2 \cdot \left\{ \left[\frac{\sin \pi \left(1 - \frac{4.8\ M}{30\ M} \right)}{\pi \left(1 - \frac{4.8\ M}{30\ M} \right)} \right]^2 \cdot \left[1 + \left(\frac{30\ M - 4.8\ M}{4.77\ M} \right)^2 \right]^{-1} + \left[\frac{\sin \pi \left(1 + \frac{4.8\ M}{30\ M} \right)}{\pi \left(1 + \frac{4.8\ M}{30\ M} \right)} \right]^2 \cdot \left[1 + \left(\frac{30\ M + 4.8\ M}{4.77\ M} \right)^2 \right]^{-1} \right\}} \right]^{-1/2} \cdot 100\%$$

$$= \left[\frac{1}{\left(\frac{0.482}{2.636} \right)^2 \cdot (0.034) + \left(\frac{-0.482}{3.644} \right)^2 \cdot (0.018)} \right]^{-1/2} \cdot 100\%$$

$$= 3.74\%FS \qquad \text{5-bit interpolated video}$$

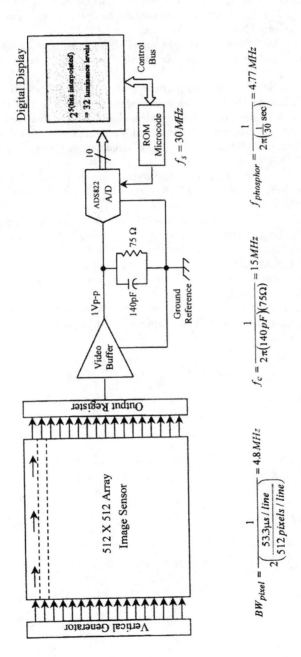

Figure 6.16 Video data conversion and recovery.

$$BW_{pixel} = \frac{1}{2\left(\dfrac{53.3\mu s / line}{512\, pixels / line}\right)} = 4.8\, MHz$$

$$f_c = \frac{1}{2\pi(140pF)(75\Omega)} = 15\, MHz$$

$$f_{phosphor} = \frac{1}{2\pi\left(\dfrac{1}{30}\, sec\right)} = 4.77\, MHz$$

Bibliography

1. Bennett, W.R. and Davey, J.R., *Data Transmission*, McGraw-Hill, New York, 1965.
2. Davenport, W.B., Jr. and Root, W.L., *An Introduction to the Theory of Random Signals and Noise*, McGraw-Hill, New York, 1958.
3. Gardenshire, L.W., "Selecting Sample Rates," *ISA Journal*, April 1964 p. 59.
4. Garrett, P.H., *Multisensor Instrumentation Six-Sigma Design*, Wiley, New York, 2002.
5. Jerri, A.J., "The Shannon Sampling Theorem — Its Various Extensions and Applications: A Tutorial Review," *Proceedings of the IEEE*, 65(11), 1977 p. 1655.
6. Kolmogoroff, A., "Interpolation and Extrapolation von Stationaren Zufalligen Folgen," *Bulletin of Academic Sciences*, Serial Mathematics, 5, 1941 (USSR).
7. Nyquist, H., "Certain Topics in Telegraph Transmission Theory," *Transactions of the AIEE*, 47, 1928 p. 617.
8. Peled, U., A Design Method with Application to Prefilters and Sampling-Rate Selection in Digital Flight Control Systems, Dissertation, Stanford University, CA, May 1978.
9. Raemer, H.R., *Statistical Communication Theory and Applications*, Prentice Hall, Englewood Cliffs, NJ, 1969.
10. Schwartz, M., Bennett, W.R., and Stein, S., *Communications Systems and Techniques*, McGraw-Hill, New York, 1966.
11. Shannon, C.E. and Weaver, W., *The Mathematical Theory of Communication*, University of Illinois Press, Urbana, 1949.
12. Wiener, N., *Extrapolation, Interpolation, and Smoothing of Stationary Time Series with Engineering Applications*, MIT Press, Cambridge, MA, 1949.

chapter seven

Advanced instrumentation systems and error analysis

7.0 Introduction

This chapter describes multisensor architectures encountered from industrial automation to analytical measurement applications. Within these information structures data are typically not fused, but instead are nonredundantly integrated to achieve improved attribution and feature characterization than is available from single sensors.

The preceding chapters have demonstrated comprehensive modeling of instrumentation systems from sensor data acquisition through signal conditioning and data conversion functions and, where appropriate, output signal reconstruction and actuation. These system models beneficially provide a physical description of instrumentation performance with regard to device and system choices to verify fulfillment of measurement accuracy. Total instrumentation error is expressed as the sum of mean error contributions plus the 1σ root sum square (RSS) of systematic and random error variances scaled as a percentage of full-scale amplitude. Modeled instrumentation system error, therefore, valuably permits performance to be quantitatively predicted *a priori*, which is especially of interest for guiding processing apparatus design to achieve sensed process observations of interest.

7.1 Integrated instrumentation design

Computer-integrated instrumentation is widely employed to interface analog measurement signals to digital systems that commonly involve joint input/output (I/O) operations, where analog signals are also recovered for actuation or end-use purposes following digital processing. Instrumentation error models derived for devices and transfer functions in the preceding chapters are presently assembled into a complete instrumentation design and I/O system analysis. This example demonstrates evaluating the cumulative error of conditioned and converted sensor signals input to a computer data bus, including their output reconstruction in analog form, with the option for substituting alternative circuit topologies and devices for further system optimization. That is especially of value for appraising different I/O products for implementation selection.

Figure 7.1 describes a high-end I/O system combining the signal conditioning analysis accompanying Figure 4.6 with Datel data converter devices to interface a digital bandpass filter for spectral resolution of amplitude signals. Signal conditioning includes a premium performance acquisition channel consisting of a 0.1%FS systematic error piezoresistive bridge–strain–gauge accelerometer that is biased by isolated ±0.5 V dc regulated excitation and connected differentially to an Analog Devices AD624C preamplifier accompanied by up to 1 V rms of common-mode random noise. The harmonic sensor signal has a maximum amplitude of 70 mV rms, corresponding to ±10 g, up to 100 Hz fundamental frequencies with a first-order rolloff to 7 mV rms at 1 kHz bandwidth. The preamplifier differential gain of 50 raises this signal to a $5V_{pp}$ full-scale value while attenuating random interference, in concert with the presampling filter, to 0.006%FS signal quality. The associated sensor-loop internal noise of 15 μV_{pp} plus preamplifier referred-to-input errors total 27 μV dc, with reference to Table 4.4. This defines an input dynamic range of $\sqrt{2}$ • 70 mV/27 μV, or 71 voltage dB, approximating 12 bits of amplitude resolution. Amplitude resolution is not further limited by subsequent system devices, which actually exceed this performance, such as the 16-bit data converters.

It is notable that the Butterworth lowpass signal-conditioning filter achieves signal quality upgrading for random noise through a linear-filter approximation to matched-filter efficiency by the provisions of equation (4.16). This filter also provides undersampled noise aliasing attenuation, described in Chapter 6, following instrumentation amplifier Av_{cm} noise rejection of V_{cm} to a negligible 0.000004%FS. Errors associated with the amplifiers, sample-hold, AMUX, analog-to-digital, and digital-to-analog data converters are primarily nonlinearities and temperature-drift contributions that result in LSB equivalents between 12 and 15 bits of accuracy. The A/D and D/A converters are also elected to be discrete-switching devices to avoid signal artifacts possible with sigma-delta type converters. Sample rate f_s, determined by dividing the available 250 kHz DMA transfer rate between eight channels, is 31 times the 1-kHz signal bandwidth. That provides excellent sampled-data performance in terms of small sinc error, negligible noise aliasing from the conditioned residual random interference, and accurate output signal recovery.

Figure 7.2 shows the error of converted input signal vs. frequency applied to a digital data bus, where its zero-order-hold intersample error value is the dominant contributor of 0.63%FS at full bandwidth. The combined total input error of 0.83%FS remains constant from 10% of signal bandwidth to the 1-kHz full-bandwidth value, owing to harmonic signal amplitude rolloff with increasing frequency, declining to 0.32%FS at 1%

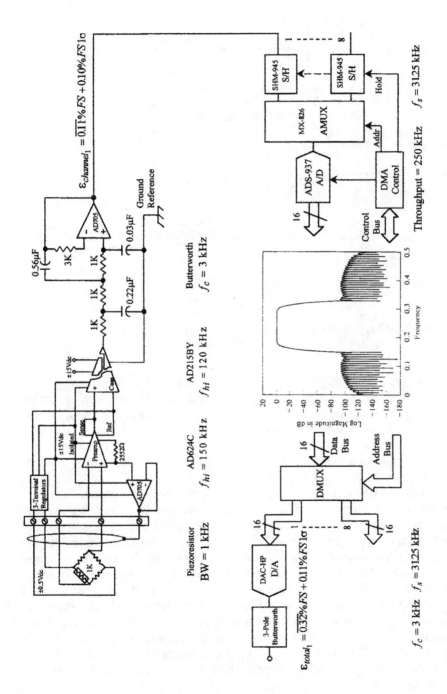

Figure 7.1 Integrated I/O instrumentation.

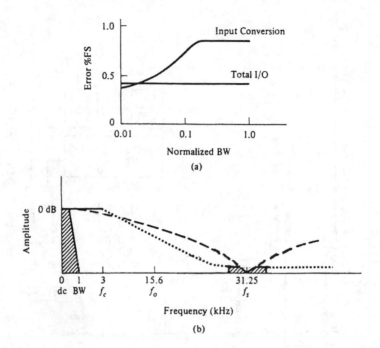

Figure 7.2 I/O system error and spectra.

bandwidth. It is significant that the sampled-image-frequency spectra described in Figure 6.13 are regenerated by each I/O sampling operation from S/H through D/A converter devices and that these spectra are transformed with signal transfer from device to device when there is a change in f_s. Increasing f_s accordingly results both in sampled-image-frequency spectra being heterodyned to higher frequencies and a decreased mean signal attenuation from the associated sinc function.

The illustrated I/O system and its accompanying models provide detailed accountability of total system performance and realize the end-to-end goal of not exceeding 0.1%FS error for any contributing element to the error summary of Table 7.1. Output signal reconstruction is effectively performed by a post-D/A Butterworth third-order lowpass filter derated to reduce its component error while simultaneously minimizing intersample error. This implementation results in an ideal flat total instrumentation channel error vs. bandwidth of 0.43%FS, shown in Figure 7.2a. This error is equivalent to approximately 8 bits of amplitude accuracy within 12 bits of signal dynamic range and 16 bits of data quantization. The output signal V_s is one-tenth of the 10-V full-scale amplitude value to represent the rolloff of signal spectra at full bandwidth by Figure 7.2b.

Table 7.1 Integrated Instrumentation Error Summary

Element	$\varepsilon_{\%FS}$	Comment
Sensor	0.100000	Piezoresistor bridge
Interface	0.010000	Excitation stability
Amplifiers	0.033600	Preamp + isolator
Presampling filter	0.115000	3-pole from Table 6.1
Signal quality	0.006020	Residual random noise
Sample-hold	0.020630	400-nsec acquisition
Analog multiplexer	0.004000	Transfer error
Noise aliasing	0.000004	Heterodyned residual
Sinc	0.084178	Sampling attenuation
A/D	0.002463	16-bit subranging
D/A	0.004280	16-bit R-2R type
Interpolator filter	0.115000	3-pole from Table 3.6
Intersample	0.000407	Output interpolation
	0.318178	$\sum \overline{mean}$
ε_{total}	0.108232	1σRSS
	0.426400	$\sum \overline{mean} + 1\sigma$RSS

Sensor

1-kΩ piezoresistor bridge with F = mA response at 0.1%FS error to KHz BW

Interface

Sensor voltage excitation ± variation ±0.5 V DC ± 50 µV or 0.01%FS

Signal Quality

$$\varepsilon_{random} = \frac{V_{cm}}{V_{diff}} \bullet \left[\frac{R_{diff}}{R_{cm}}\right]^{1/2} \bullet \frac{Av_{cm}}{Av_{diff}} \bullet \left[\frac{2}{k}\frac{f_c}{f_{hi}}\right]^{1/2} \bullet 100\%$$

$$= \frac{1\,V}{7\,mV} \bullet \left[\frac{1\,G\Omega}{1\,G\Omega}\right]^{1/2} \bullet \frac{10^{-4}}{50} \bullet \left[\frac{2}{0.9}\frac{3\,kHz}{150\,kHz}\right]^{1/2} \bullet 100\%$$

$$= 0.006020\%FS$$

Sample-Hold

Acquisition error	0.00760%
Nonlinearity	0.00040
Gain	0.02000
Temperature coefficient	0.00500
$\varepsilon_{S/H}$	0.020630%FS 1σRSS

Amplifiers		
Parameter	AD624C	AD215BY
V_{os}	Trimmed	Trimmed
$\dfrac{dV_{os}}{dT} \bullet dT$	2.5 µV	20 µV
$I_{os} \bullet R_i$	$\overline{10\ \mu V}$	$\overline{15\ \mu V}$
$V_{N_{pp}}$	15 µV	2 µV
$f(\text{Av}) \bullet \dfrac{Vo_{FS}}{\text{Av}_{diff}}$	$\overline{1\ \mu V}$	$\overline{250\ \mu V}$
$\dfrac{d\text{Av}}{dT} \bullet dT \bullet \dfrac{Vo_{FS}}{\text{Av}_{diff}}$	5 µV	750 µV
$\varepsilon_{\text{ampl RTI}}$	$(\overline{11}+16)\mu V$	$(\overline{265}+750)\mu V\ \sum\overline{mean}\ +1\sigma\text{RSS}$
$\varepsilon_{\text{ampl \%FS}}$	0.027%FS	0.020%FS $\times \dfrac{Vo_{FS}}{\text{Av}_{diff}} \bullet 100\%$

Analog Multiplexer	
Transfer error	$\overline{0.003\%}$
Leakage	0.001
Crosstalk	0.00005
$\varepsilon_{\text{AMUX}}$	$\overline{0.0040\%}$FS $\qquad \sum\overline{mean}\ +1\sigma\text{RSS}$

Noise Aliasing

$$\varepsilon_{\text{random alias}} = \frac{\sqrt{2}\ 100\%}{[\text{SNR}_{\text{random alias}}]^{1/2}}$$

$$= \frac{\sqrt{2}\ 100\%}{\left[V_{FS}^2/(V_{\text{noise rms}})^2 \bullet \left(\cfrac{1}{\left[1+\left(\dfrac{f_s}{f_c}\right)^{2n} \right]^{1/2}} \right)^2 \right]^{1/2}}$$

$$= \frac{\sqrt{2}\ 100\%}{\left[V_{FS}^2 \Big/ \left(\frac{\sqrt{2} \bullet 1 V_{rms} \bullet 10^{-4}}{5V} V_{FS} \right)^2 \bullet \left(\frac{1}{\left[1 + \left(\frac{31.25\ \text{kHz}}{3\ \text{kHz}} \right)^6 \right]^{1/2}} \right)^2 \right]^{1/2}}$$

$$= \frac{\sqrt{2}\ 100\%}{\left[V_{FS}^2 \Big/ \left(9 \times 10^{-10} V_{FS}^2 \right)(0.78 \times 10^{-6}) V_{FS}^2 \right]^{1/2}}$$

$$= \frac{\sqrt{2}\ 100\%}{\left[V_{FS}^2 / 7 \times 10^{-16} V_{FS}^2 \right]^{1/2}}$$

$$= 0.000004\%\text{FS}$$

Sinc

$$\varepsilon_{sinc} = \frac{1}{2} \left(1 - \frac{\sin \pi\ BW/f_s}{\pi\ BW/f_s} \right) \bullet 100\%$$

$$= \frac{1}{2} \left(1 - \frac{\sin \pi\ 1\ \text{kHz}/31.25\ \text{kHz}}{\pi\ 1\ \text{kHz}/31.25\ \text{kHz}} \right) \bullet 100\%$$

$$= \overline{0.084178\%\text{FS}}$$

16-Bit A/D

Mean integral nonlinearity (1 LSB)	0.0011%
Noise + distortion	0.0001
Quantizing uncertainty (1/2 LSB)	0.0008
Temperature coefficient	0.0011
$\varepsilon_{A/D}$	0.002463%FS Σ_{mean} + 1σRSS

16-Bit D/A

Mean integral nonlinearity (1 LSB)	0.003%
Noise + distortion	0.0008
Temperature coefficient	0.0010
$\varepsilon_{D/A}$	0.004280%FS Σ_{mean} + 1σRSS

Interpolated Intersample

$$\varepsilon_{\Delta V} = \left[\frac{Vo_{\mathrm{FS}}^2}{V_s^2 \left\{ \sin c^2 \left(1 - \dfrac{\mathrm{BW}}{f_s} \right) \left[1 + \left(\dfrac{f_s - \mathrm{BW}}{f_c} \right)^{2n} \right]^{-1} + \sin c^2 \left(1 + \dfrac{\mathrm{BW}}{f_s} \right) \left[1 + \left(\dfrac{f_s + \mathrm{BW}}{f_c} \right)^{2n} \right]^{-1} \right\}} \right]^{-1/2} \bullet 100\%$$

$$= \left[\frac{(10V)^2}{(1V)^2 \bullet \left\{ \sin c^2 \left(1 - \dfrac{1\,\mathrm{kHz}}{31.25\,\mathrm{kHz}} \right) \left[1 + \left(\dfrac{31.25\,\mathrm{kHz} - 1\,\mathrm{kHz}}{1\,\mathrm{kHz}} \right)^6 \right]^{-1} + \sin c^2 \left(1 + \dfrac{1\,\mathrm{kHz}}{31.25\,\mathrm{kHz}} \right) \left[1 + \left(\dfrac{31.25\,\mathrm{kHz} + 1\,\mathrm{kHz}}{1\,\mathrm{kHz}} \right)^6 \right]^{-1} \right\}} \right]^{-1/2} \bullet 100\%$$

$$= 0.000407\%\mathrm{FS}$$

7.2 Multisensor error propagation

Figure 7.3 describes a multisensor measurement process applicable to turbine engine manufacture for evaluating blade internal airflows, with respect to design requirements, essential to part heat transfer and rogue blade screening. A preferred evaluation method is to describe blade airflow in terms of fundamental geometry, such as its effective flow area. The implementation of this measurement process is described by analytical equations

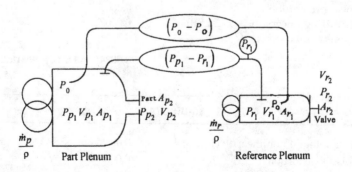

Figure 7.3 Part airflow measurement process.

Table 7.2 Airflow Process Parameter Glossary

Known Airflow Process Parameters			Measured Airflow Process Parameters		
Symbol	Value	Description	Symbol	Value	Description
$\dfrac{\dot{m}_r}{\rho}$	$\dfrac{ft^3}{min}$	Reference plenum volumetric flow	Ap_2	ft^2	Part effective flow area
Ar_1	ft^2	Reference plenum inlet area	$Pp_1 - Pr_1$	lb/ft^2	Part-to-reference plenum differential pressure
Vr_1	$\dfrac{ft}{min}$	Reference plenum inlet velocity	Pr_1	lb/ft^2	Reference plenum-gauge pressure
Ap_1	ft^2	Part plenum inlet area	$P_o - P_o$	lb/ft^2	Reference-part plenum equalized stagnation pressures
ρ	0.697E-6 $\dfrac{lb-min}{ft^4}$	Air density at standard temperature and pressure	Pp_2	$P_{atm}lb/ft^2$	Part plenum exit pressure

(7.1) and (7.2), where uncontrolled air density, ρ, appears as a ratio to effect an air-density-independent airflow measurement. That outcome beneficially enables quantitative determination of part airflow area from known parameters and parameter measurements that are defined in Table 7.2. The airflow process mechanization accordingly consists of two plenums with specific volumetric airflows and four pressure measurements. Appendix A describes the derivation of analytical equations on which this process is based.

In operation, the fixed and measured quantities determine part flow area, employing two measurement sequences. Plenum volumetric airflows are initially reconciled for pitot stagnation pressures P_o-P_o, thereby obtaining the plenums ratio of internal airflow velocities, V_{p1}/V_{r1}. The quantities are then arranged into a ratio of plenum volumetric airflows which, combined with gauge and differential pressure measurements P_{r1}, P_{atm}, and P_{p1}-P_{r1}, permit expression of air density-independent part flow area A_{P2} by equation (7.2). Equationl (7.3) then describes multisensor error propagation determined from analytical process equations (7.1) and (7.2) with the aid of Table 7.3. Part flow area error results from the algorithmic propagation of four independent pressure sensor instrumentation errors in this two-sequence measurement algorithm, where individual sequence errors are summed because of the absence of correlation between the measurements.

Table 7.3 Instrumentation Error Algorithmic Propagation

Instrumentation Error	Algorithmic Operation	Error Influence
	Addition	$\sum \overline{\varepsilon_{mean}}$ %FS
	Subtraction	$\sum \overline{\varepsilon_{mean}}$ %FS
$\varepsilon_{mean\,\%FS}$	Multiplication	$\sum \overline{\varepsilon_{mean}}$ %FS
	Division	$\sum \overline{\varepsilon_{mean}}$ %FS
	Power function	$\sum \overline{\varepsilon_{mean}}$ %FS $\times$ |exponent value|
	Addition	1σRSS ε %FS
	Subtraction	1σRSS ε %FS
$\varepsilon_{\%FS\,1\sigma}$	Multiplication	1σRSS ε %FS
	Division	1σRSS ε %FS
	Power function	1σRSS ε %FS $\times$ |exponent value|

In the first sequence an equalized pitot pressure measurement, ΔP_o, is acquired, defining Bernoulli's equation (7.1). The algorithmic influence of this pressure measurement is represented by the sum of its static mean plus single RSS error contribution in the first sequence of equation (7.3). The second measurement sequence is defined by equation (7.2), whose algorithmic error propagation is obtained from arithmetic operations on measurements P_{r1}, P_{atm}, and P_{p1}-P_{r1} represented by the sum of their mean plus RSS error contributions in equation (7.3). For the first sequence of equation (7.3), only the differential-pitot-stagnation-pressure measurement P_o-P_o is propagated as algorithmic error. In the following second sequence, part plenum inlet area, A_{p1}, air density, ρ, and reference plenum inlet velocity, V_{r1}, are constants that do not appear as propagated error. However, the square-root exponent influences the error of the three pressure measurements included in equation (7.2) by the absolute value of $|1/2|$ shown. Four independent 9-bit accuracy pressure measurements are accordingly combined by these equations to realize an 8-bit accuracy part flow area. Equation (7.4) defines the error for each pressure measurement, ε_{sensor}, including its signal conditioning and data conversion at the host computer data bus by the methods of the preceding chapters.

$$\Delta P_o = \left(P_{p1} + 1/2\rho V_{p1}^2 \right) - \left(P_{r1} + 1/2\rho V_{r1}^2 \right) \qquad \text{equilibrium sequence} \qquad (7.1)$$

$$A_{P2} = A_{P1} \bullet \left[\frac{\rho - 2(P_{p1} - P_{r1})/V_{r1}^2}{\rho + 2(P_{r1} - P_{atm})/V_{r1}^2} \right]^{1/2} \qquad \text{part flow area sequence} \qquad (7.2)$$

$$\varepsilon_{\Delta P_0} + \varepsilon_{A_{P2}} = \left\{ \overline{\varepsilon_{\text{mean } \Delta P_o}} \, \%FS + \varepsilon_{\Delta P_o} \, \%FS1\sigma \right\}_{\text{1st sequence}} \qquad \text{error propagation} \qquad (7.3)$$

$$+ \left\{ \left| \frac{1}{2} \right| \left[\overline{\varepsilon_{\text{mean } \Delta P_{p1\text{-}r1}}} + \overline{\varepsilon_{\text{mean } P_{r1}}} + \overline{\varepsilon_{\text{mean } P_{\text{atm}}}} \right] \%FS \right.$$

$$\left. + \left| \frac{1}{2} \right| \left[\varepsilon^2_{\Delta P_{p1\text{-}r1}} + \varepsilon^2_{P_{r1}} + \varepsilon^2_{P_{\text{atm}}} \right] \%FS1\sigma \right\}_{\text{2nd sequence}}$$

$$= \left\{ \overline{0.1\%} \, FS + 0.1\%FS \, 1\sigma \right\}_{\text{1st sequence}}$$

$$+ \left\{ \left| \frac{1}{2} \right| \left[\overline{0.1} + \overline{0.1} + \overline{0.1} \right] \%FS \right.$$

$$\left. + \left| \frac{1}{2} \right| \left[0.1^2 + 0.1^2 + 0.1^2 \right]^{1/2} \%FS \, 1\sigma \right\}_{\text{2nd sequence}}$$

$$= \overline{0.25} \, \%FS + 0.186\%FS \, 1\sigma \quad \text{8-bit accuracy}$$

$$\varepsilon_{\text{sensor}} = \overline{0.1} \, \%FS + 0.1\%FS1s \quad \text{data acquisition channel} \qquad (7.4)$$

7.3 Robotic axes volumetric error

An alternative system error is encountered in the determination of absolute position within the volume of multiaxis robotic and dimensional coordinate measurement machines (CMMs) whose evaluation is typically a formidable task. For large machines, such as encountered in medical applications and industrial processes, achieving high accuracy is especially difficult because of machine apparatus size and heavy payloads or workpieces. Further, in manufacturing enterprises the versatility of CAD systems coupled to NC (numerically controlled) machine tools have increased productivity while increasing the need for machine apparatus dimensional accuracy. This has arisen to meet advancing product quality requirements in the translation of ideal three-dimensional information into actual product realizations, introduced here as combined volumetric error defined by ideal-to-actual product geometry.

Geometric error is attributable to deviations in the structural accuracy of a machine. A contemporary evaluation method employed for machine volumetric error resolution is laser interferometry, with available mean measurement error to 1 μm at a 1σ coordinate uncertainty to one-tenth of a micrometer. Five sample volumetric component errors are shown in Figure 7.4, including both mean and uncertainty contributions. Their XYZϕ combined volumetric error is described by Table 7.4.

The CMM geometric errors include X-axis table mean squareness error, relative to the Y-axis and Z-axis perpendicular plane, which is defined relative to a vertical flatness parallelepiped as shown. X-axis table mean straightness

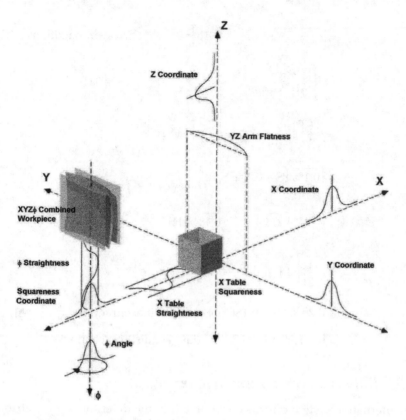

Figure 7.4 Coordinate measurement machine errors.

error is projected in a plane bounded by two parallel lines. Y-axis and Z-axis mean flatness error of a movable arm is defined by deviation from the ideal (dashed) flatness vertical plane. In contrast, φ-axis mean straightness error defines workpiece perpendicularity vertically from the X-axis table projected in a plane bounded by two vertical lines. Accompanying XYZφ coordinate

Table 7.4 Volumetric Mean and Uncertainty Error

Component	Axis	Mean Error	Uncertainty
Table	X	Squareness	Coordinate
		Straightness	Coordinate
Movable arm	Y	Flatness	Coordinate
	Z	Flatness	Coordinate
Workpiece	φ	Straightness	Coordinate
Combined	XYZφ	$\Sigma(\overline{mean})$ +	1σRSS uncertainties

uncertainties define the statistical variability in each of the five volumetric component errors. A rotary table CMM is common for accommodating multifaceted and sculptured workpieces, as shown.

Bibliography

1. Brockman, J.P., Error Referenced Non-Redundant Sample Coding for Data Conversion Systems, Master's Thesis, Electrical Engineering, University of Cincinnati, 1986.
2. Droder, G.R., Performance Adaptive Error Control for Real-Time Computer I/O Applications in Integrated Manufacturing Systems, Master's Thesis, Electrical Engineering, University of Cincinnati, 1985.
3. Garrett, P.H., *Advanced Instrumentation and Computer I/O Design*, IEEE Press, 1994.
4. Garrett, P.H., *Analog I/O Design: Acquisition-Conversion-Recovery*, Reston Publishing Co., Reston, VA, 1981.
5. Garrett, P.H., *Multisensor Instrumentation 6σ Design*, Wiley, New York, 2002.
6. Gordon, B.M., *The Analogic Data Conversion Systems Digest*, Analogic, Wakefield, MA, 1977, 2nd edition.
7. Katz, P., *Digital Control Using Microprocessors*, Prentice Hall, New York, 1981.
8. Kester, W., "Test Video A/D Converters under Dynamic Conditions," *Electronic Design News*, August 18, 1982 p. 103.
9. Peled, U, A Design Method with Application to Prefilters and Sampling Rate Selection in Digital Flight Control Systems, Aeronautics Dissertation, Stanford University, CA, May 1978.
10. Wintz, P., and Gonzales, R.C., *Digital Image Processing*, Addison-Wesley, Reading, MA, 1977.
11. Zuch, E.L., *Data Acquisition and Conversion Handbook*, Datel-Intersil, Mansfield, MA, 1979.

Appendix A

Lemma 1
Continuity equation for the reference plenum provides inlet Vr_1:

$$m_r = \rho Ar_1 Vr_1 = \rho Ar_2 Vr_2 \qquad \text{reference plenum}$$

$$m_p = \rho Ap_1 Vp_1 = \rho Ap_2 Vp_2 \qquad \text{part plenum}$$

$$\therefore Vr_1 = \frac{m_r/\rho}{Ar_1} = \frac{25 \text{ ft}^3/\text{min}}{0.1724 \text{ ft}^2} = 145 \text{ ft/min}$$

Lemma 2
Adjusting m_p/ρ equalizes plenums stagnation, P_o, for inlet $\dfrac{Vp_1}{Vr_1}$:

$$P_o = Pp_1 + \tfrac{1}{2}\rho \, Vp_1{}^2 \qquad \text{total part pressure}$$
$$P_o = Pr_1 + \tfrac{1}{2}\rho \, Vr_1{}^2 \qquad \text{total reference pressure}$$

$$\therefore Pp_1 - Pr_1 + \tfrac{1}{2}\rho\left(Vp_1^2 - Vr_1^2\right) = 0 \qquad \text{for } P_o \text{ equalized}$$

$$\frac{Vp_1}{Vr_1} = \sqrt{1 - \frac{2(Pp_1 - Pr_1)}{\rho Vr_1^2}} \qquad \text{velocity ratio of plenums}$$

Lemma 3
Bernoulli equation determines part exit Vp_2:

$$P_o = Pp_2 + \tfrac{1}{2}\rho \, Vp_2^2 \qquad P_o \text{ constant throughout plenum}$$

$$Vp_2 = \sqrt{\frac{2(P_o - P_{\text{atm}})}{\rho}} \qquad \text{where part } Pp_2 = P_{\text{atm}}$$

Lemma 4
Exit $\dfrac{Vp_2}{Vr_2}$ and continuity equations offer identity for $\dfrac{\dot{m}_p}{\dot{m}_r}$:

$$\frac{Vp_2}{Vr_2} = \frac{Vp_1}{Vr_1} \bullet \frac{Ap_1}{Ap_2} \bullet \frac{Ar_2}{Ar_1}$$

$$= \frac{m_p/\rho}{m_r/\rho} \bullet \frac{Ar_2}{Ap_2} \qquad \text{continuity equation identity}$$

$$\therefore \; \frac{m_p/\rho}{m_r/\rho} = \frac{Ap_1}{Ar_1} \bullet \sqrt{1 - \frac{2(Pp_1 - Pr_1)}{\rho Vr_1^2}}$$

Lemma 5

Continuity equation defines part flow area Ap_2 rationalized by blower ratio:

$$Ap_2 = \frac{m_p}{\rho} \bullet \frac{1}{Vp_2} \bullet \frac{m_r/\rho}{m_r/\rho}$$

$$= \frac{m_p}{\rho} \bullet \frac{1}{\sqrt{\dfrac{2\left[\left(1/2\rho Vr_1^2 + Pr_1\right) - P_{atm}\right]}{\rho}}} \bullet \frac{m_r/\rho}{m_r/\rho}$$

$$= \frac{m_p}{m_r} \bullet \frac{m_r}{\rho} \bullet \frac{1}{\sqrt{Vr_1^2 + \dfrac{2(Pr_1 - P_{atm})}{\rho}}}$$

Lemma 6

Air density independent part flow area:

$$Ap_2 = \frac{Ap_1}{Ar_1} \bullet \sqrt{1 - \frac{2(Pp_1 - Pr_1)}{\rho Vr_1^2}} \bullet \frac{m_r}{\rho} \bullet \sqrt{Vr_1^2 + \frac{2(Pr_1 - P_{atm})}{\rho}}$$

$$= \frac{Ap_1}{Ar_1} \bullet \frac{m_r/\rho}{Vr_1} \bullet \sqrt{\frac{\rho - \dfrac{2(Pp_1 - Pr_1)}{Vr_1^2}}{\rho + \dfrac{2(Pr_1 - P_{atm})}{Vr_1^2}}}$$

$$= Ap_1 \bullet \sqrt{\frac{\rho - 2(Pp_1 - Pr_1)/Vr_1^2}{\rho + 2(Pr_1 - P_{atm})/Vr_1^2}} \qquad \text{part flow area}$$

Lemma 7

Open part plenum limit proof for Ap_2:

$$Ap_2 = Ap_1 \bullet \sqrt{\frac{\rho - 2(P_{atm} - Pr_1)/Vr_1^2}{\rho + 2(Pr_1 - P_{atm})/Vr_1^2}} \qquad Pp_1 = P_{atm}$$

$$= Ap_1 \bullet \sqrt{\frac{\rho + 2(Pr_1 - P_{atm})/Vr_1^2}{\rho + 2(Pr_1 - P_{atm})/Vr_1^2}} = Ap_1$$

chapter eight

Automation systems concurrent engineering

8.0 Introduction

Manufacturing science has a legacy of innovation, leading to commercialization of high-value products ranging from semiconductors and composite materials to superconductors and metals. Advancement results from the search for solutions to significant challenges such as the molecular processing of materials presented. Early process designs often were treated as prescriptive combinatorial systems absent processing models and with only postprocessing product characterization. Improved insight into process understanding subsequently evolved to a refinement of sensing multiple internal product properties during processing for the automation of mass, momentum, and energy quantities employing focused subprocess control. Concurrent engineering is shown to offer a systematic framework of process design for production by integrating the best practices for obtaining robust manufacturing performance and product value.

8.1 Concurrent engineering common core

With reference to Figure 8.1, six generic process control linkages are developed in this chapter. The first linkage describes how product manufacture is enabled by the identification of product property physical apparatus variables. However, the often disparate translation between product properties and apparatus variables requires additional capabilities of process automation systems encompassed by attributes of the concurrent engineering linkages. One such attribute is robust process design to achieve axiomatic apparatus multivariable decoupling for reduction of subprocess interactions. This is described by the second linkage. Manufacturing processes further require the direction of multiple controlled variables along ideal state trajectories with a feedforward-modeled *ex situ* planner by linkage three. The planner defines ideal process behavior throughout a production cycle including the attenuation of unmodeled processing dynamics, and often employs

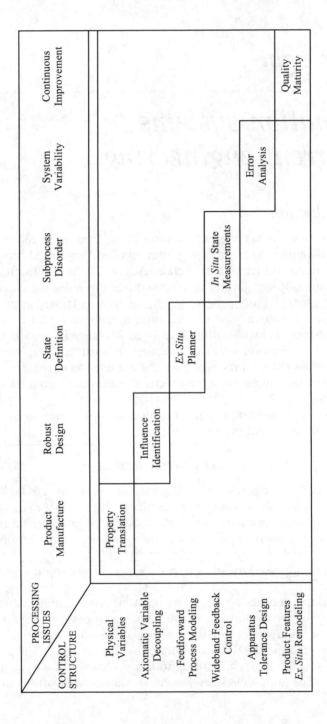

Figure 8.1 Process automation concurrent engineering.

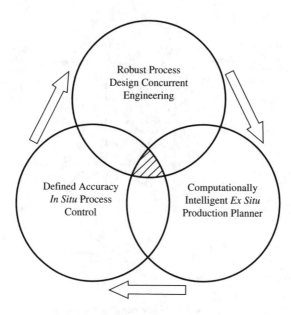

Figure 8.2 Automation systems quality deployment.

computational intelligence in its realization when quantitative models are inadequate.

The fourth linkage details the implementation of multisensor *in situ* subprocess wideband feedback control for regulation of mass, momentum, and energy quantities. Meeting *in situ* subprocess response requirements with equivalent control bandwidth is a principal distinction of *in situ* processing. Mechanization typically involves offline analytical product data, online property or reaction *in situ* measurements, and environmental parameter gauging. Linkage five provides for system uncertainty analysis whereby the narrowest information and apparatus tolerance designs are reserved for the largest contributing uncertainty components and include the instrumentation error modeling methods developed in the preceding chapters. Linkage six enables a pathway to product quality maturity by process data mining for continuous product improvement. Quality is defined as consistent product property conformance to goal specifications. Sensor-based product feature assessment is demonstrated for enhanced product processing by means of online *ex situ* planner remodeling. Process automation performance is accordingly advanced by these concurrent engineering process and control design linkages, as captured by the quality deployment of Figure 8.2.

8.2 *Product property apparatus variables*

Product manufacture requires property models with physical interpretations that identify apparatus variables and their actuation with sufficient resolution and dynamic range, in order to describe processing sequences capable

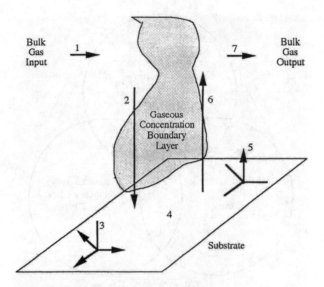

1. Forced Flow of Reactant Gases
2. Diffusion and Bulk Flow of Reactants through Gaseous Layer
3. Adsorption of Reactants onto Surface
4. Deposition and other Chemical Processes on Surface
5. Desorption of Products from Surface
6. Diffusion and Bulk Flow of Products through Gaseous Layer
7. Forced Exit of Gases

Figure 8.3 Chemical vapor deposition properties.

of realizing product composition and structure goals. This knowledge is often formed by generalizations derived from domain-specific experience. In cases where quantitative models are not available, qualitative models may be described for process path generation under the aegis of intelligent process control.

Attachment of atoms to a surface at the atomic scale is a complex physical event. Nucleation of films on substrates involves a series of random interactions with competitive growth in composition and microstructure. Figure 8.3 is shown for multiple materials and layers to, for example, illustrate that thermodynamic modeling is necessary but not sufficient to successfully predict processing. Chemical deposition rate is determined by intermediate reactions at a surface as well as mass transport of the reactants. Seven deposition influences illustrate the challenges confronted in chemical vapor deposition (CVD) processes that are modeled by finite-element Navier-Stokes partial differential equations (8.1) through (8.3).

Conservation of mass:

$$v_r \frac{\partial w_i}{\partial r} + v_z \frac{\partial w_i}{\partial z} = \frac{1}{r} \frac{\partial}{\partial r} r D_i \frac{\partial w_i}{\partial r} + \frac{\partial}{\partial z} D_i \frac{\partial w_i}{\partial z} + \sum_{k=1}^{k} R_{ik} \qquad (8.1)$$

Conservation of momentum:

$$v_r \frac{\partial}{\partial r}(\rho v_z) + v_z \frac{\partial}{\partial z}(\rho v_z) = -\frac{\partial p}{\partial z} + \frac{1}{r}\frac{\partial}{\partial r}\mu r \frac{\partial v_z}{\partial r} + \frac{\partial}{\partial z}\mu \frac{\partial v_z}{\partial z} + \rho g_z \qquad (8.2)$$

Conservation of thermal energy:

$$v_r \frac{\partial}{\partial r}(\rho C_p T) + v_z \frac{\partial}{\partial z}(\rho C_p T) = \frac{1}{r}\frac{\partial}{\partial r}kr\frac{\partial T}{\partial r} + \frac{\partial}{\partial z}k\frac{\partial T}{\partial z} \qquad (8.3)$$

In the conservation of mass equation, W_i represents a mass fraction of specie i, D_i is the diffusion coefficient, and R_i is the chemical generation rate in cylindrical coordinates. For the conservation of momentum, ρ defines gas mixture density proportional to total pressure p, v_z and v_r represent reactant gas velocities in axial and radial directions, μ is the gas molecular viscosity, and the ρg_z product refers to convective flow. These enable accurate prediction of gas concentrations and deposition profiles in a reactor but not real-time control because of their computationally intensive solution requirement.

The Navier-Stokes equations, however, do reveal essential variables for measurement and manipulation purposes. Figure 8.4 illustrates available apparatus, sensors, and actuators that alone are capable of only approximating required product properties, which underscores the value of the compensatory concurrent engineering methods shown in Figure 8.1. Liquid precursors further assist precise process control of stoichiometry aided by reactant gas mass flow controllers, including gas-stream temperature regulation by oxygen addition and *in situ* chemical mass spectrometry compared with final product *ex situ* chemical Raman spectrometry. With this structure product properties (chemistry, morphology) can be beneficially influenced by actuation variables (heat, pressure, flow) to achieve the highest growth process states (mass transfer reactions) occurring for *in situ* gas-stream temperatures that maintain deposition in the mass-transport CVD processing region.

8.3 Robust process axiomatic design

Concurrent engineering provides for the joint consideration of product manufacture and its process design requirements. For product realizations it is beneficial for mapping between product functional properties and corresponding apparatus physical variables to be either uncoupled or decoupled, where an equivalent matrix mapping is, respectively, either diagonal or triangular. The subprocess architecture aids this decomposition by revealing more precise modeling regions, partitioning for parameter linearity, and defining intersubprocess variable influences. The degree of subprocess decomposition required depends on both the achievable modeling granularity and the control difficulty confronted. Figure 8.5 describes three automation levels, including environmental, *in situ*, and product with a nested process representation.

Robust product processing necessitates the accommodation of multiple apparatus variables with accuracy, reliability, and minimum disorder. These

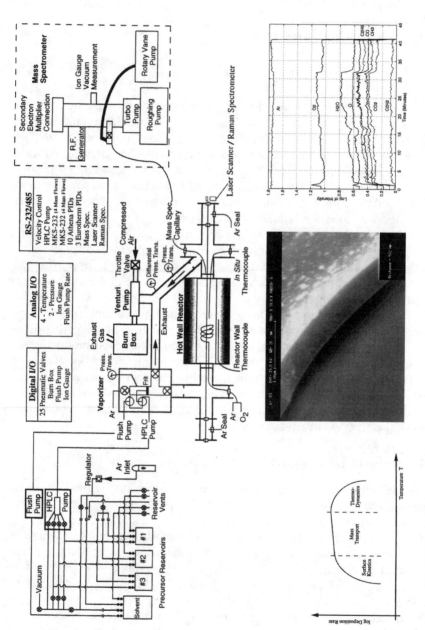

Figure 8.4 CVD product processing apparatus variables.

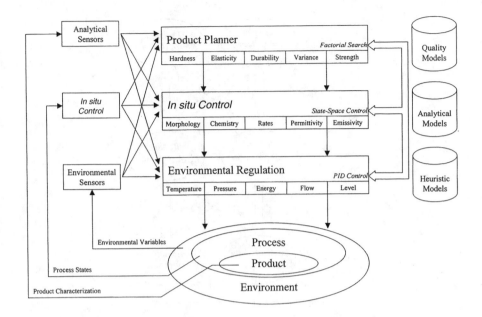

Figure 8.5 Hierarchical process automation reference architecture.

qualities are enabled by axiomatic process design incorporating intersubprocess variable decoupling, evidenced by decreased deviation from ideal state values during processing. Process design and control knowledge requirements are also correspondingly less with uncoupled or decoupled subprocess variables because of the minimization of bias, processing disorder, and reduced control iteration inefficiencies. That achievement corresponds to knowledge focusing in fewer consecutive processing states for increased performance effectiveness.

Figure 8.6 illustrates ideal matrix-diagonal-only uncoupled subprocess-to-subprocess variables determining processing outcomes by solid lines and crosscoupled disorder-inducing matrix-off-diagonal covariance influences by dashed lines. Axiomatic influence design accordingly is introduced for multivariable processes in contrast to per-variable algorithmic design. In practice, subprocess variables are synthesized or restructured by process apparatus configuration to obtain rectangular parameter matrices that are at least decoupled, as defined by achieving as completely as possible zero covariance parameters on one side of the subprocess mapping matrix diagonals shown in equation (8.4). The illustrated product, *in situ*, and environmental subprocess matrices must all be either upper triangular or lower triangular to achieve decoupled effectiveness.

$$\begin{bmatrix} \delta \\ \vdots \\ \mu \end{bmatrix} = \begin{bmatrix} P_{11} & \cdots & \\ \vdots & \ddots & \\ & & P_{nn} \end{bmatrix} \begin{bmatrix} I_{11} & \cdots & \\ \vdots & \ddots & \\ & & I_{nn} \end{bmatrix} \begin{bmatrix} E_{11} & \cdots & \\ \vdots & \ddots & \\ & & E_{nn} \end{bmatrix} \begin{bmatrix} \mu_1 \\ \vdots \\ \mu_n \end{bmatrix} \quad (8.4)$$

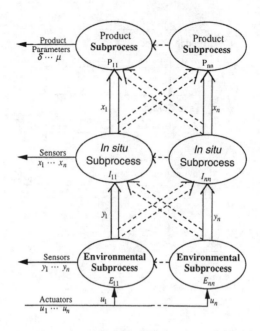

Figure 8.6 Subprocess axiomatic influences.

8.4 *Automated production* ex situ *planner*

Manufacturing processes require a foundation of production models for direction of multiple controlled variables along ideal state trajectories in order to achieve product properties of interest. An *ex situ*-planner-directed control system enables this achievement by defining process behavior throughout a processing cycle. Planner outputs accordingly provide *in situ* references, based on inputs of product characterization data and online *in situ* process and environmental apparatus sensor measurements. This is shown in Figure 8.7.

The *ex situ* feedforward model is capable of event-based process compensation beyond the capabilities of *in situ* and environmental feedback control alone. This is obtained by generating ideal *ex situ* planner *in situ* references that are differenced with actual *in situ* state measurements in the *in situ* controller and output to environmental control loops for *in situ* actuation. A principal performance advancement is attenuation of unmodeled processing dynamic events arising from both internal and external sources of process disorder. That is physically achieved by the combined effectiveness of feedforward control for reducing long-time-constant process variability and feedback control for regulating short-time-constant process variability.

In addition to quantitative *ex situ* planner models, nonprescriptive qualitative *ex situ* planners are also demonstrated with rule-based expert systems, neural networks, and fuzzy logic examples of computational intelligence in the process application examples that follow. Rule-based models utilize expert's domain knowledge understanding of process behavior, whereas

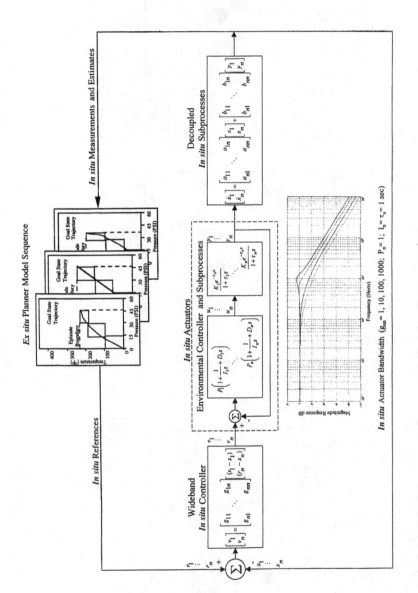

Figure 8.7 Ex situ-planner-referenced in situ subprocess control.

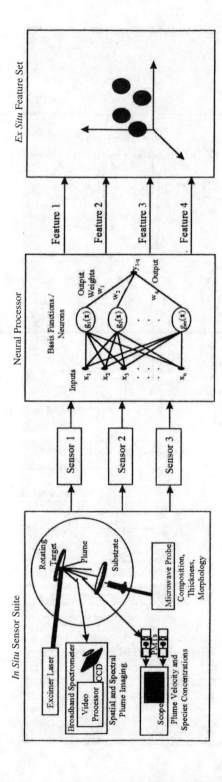

Figure 8.8 ANN features-directed *ex situ* planner sensor fusion.

neural networks provide optimum nonlinear parameter mappings trained on quantitative process data. Fuzzy logic encodes quantitative process data with a symbolic representation of actuation options for smoothing across the control domain. Figure 8.8 illustrates a neural network-assisted process automation system.

8.5 Molecular in situ subprocess control

Ex situ planner ideal references are executed over product processing paths by *in situ* controller outer feedback loops, where *in situ* actuation is constituted by inner feedback loops consisting of environmental apparatus controllers and subprocesses. This model–reference control structure is detailed by Figure 8.9 and Table 8.1, showing the hierarchical relationship between discrete subprocesses and their controllers. *In situ* controlled variables, x_n, accordingly are actuated by environmental controlled variables, y_n, which are accompanied by a reduction in environmental variable time constants as the inverse of *in situ* controller gain $\tau_n/(1+g_{nn})$, resulting from environmental-to-*in situ* subprocess cascade control. That result beneficially widens *in situ* actuation bandwidth, enabling microsecond-rate feedback control of the x_n mass–momentum–energy process state values. Meeting *in situ* subprocess response requirements with like control bandwidth is essential for effective processing, as enabled by the process control structure shown in Figure 8.7.

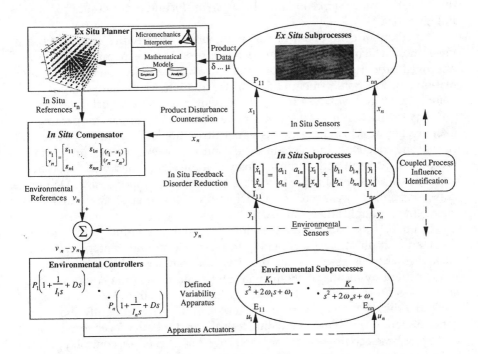

Figure 8.9 Molecular process automation influences.

Table 8.1 Hierarchical Process Automation Subprocesses

Ex situ planner	Feedforward event-based compensation of long-time-constant process variability employing analytical algorithms and computational intelligence whose output provides *in situ* controller references from inputs of product, *in situ*, and environmental data.
In situ control	Feedback control of *in situ* state product properties based on mass, momentum, and energy measurements that are differenced with *ex situ* planner references to enable greater reduction of short-time-constant process variability than environmental apparatus process control.
Environmental regulation	Ultralinear controlled variables provided by trapezoidal-tuned PID controllers and narrow tolerance component design that close environmental apparatus control loops constituting *in situ* process actuators.

It is significant that in the absence of this *in situ* subprocess response and control bandwidth equality that only environmental regulation will exist having open-loop *in situ* subprocess interactions without the benefit of *in situ* control. Functionally, *in situ* sensors provide inspection during processing for nondestructive product evaluation. Observe that the B matrix coefficients describe linkages between environmental and *in situ* subprocesses, whereas the A coefficients describe linkages between *in situ* and product subprocesses in Figure 8.9.

An essential component of this automation architecture is multisensor integration, which aggregates heterogeneous process measurements to achieve improved property characterization than is available from single sensors, for real-time product assessment and control actuation. Molecular process automation typically involves a trilogy of sensed analytical product data, *in situ* property or reaction measurements, and environmental apparatus parameter gauging. Figure 8.10 delineates this signal attribution at each subprocess level, supporting effective subprocess-focused control. However, control uncertainty is ultimately determined by the accuracy of these signals and their associated apparatus, as described in the following section on tolerance design. Descriptions of specific process sensors are deferred until their appearance in subsequent chapter application examples.

Figure 8.11 describes available control technologies employed in process automation applications that intersect the system examples that follow. Feedback regulation is essential for physical apparatus variables and is often implemented by means of commercial PID controllers at the environmental subprocess level shown in Figure 8.9. Model-based control is the primary method utilized for *ex situ* planner implementations, frequently aided by computationally intelligent modeling techniques, for both process production and event-based compensation of processing disorder. *In situ* control typically relies on state-space control representations because of their complexity

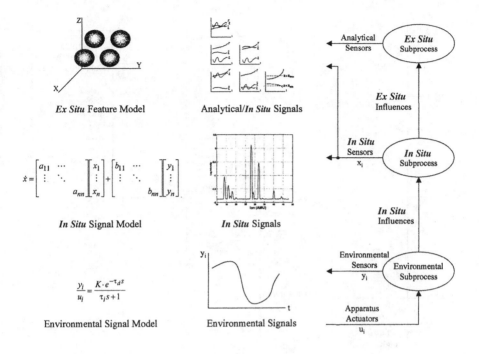

Figure 8.10 Molecular process automation signal hierarchy.

accommodation, with extension to adaptive control when process parameter migration necessitates remodeling and control compensator retuning.

8.6   System variability tolerance analysis

Process planner and knowledge-based models benefit from efficient information representations and exchanges that are compactly measurable by the entropy criterion, in the range $0 \leq H < 1$, to provide a unified evaluation of uncertainty in automation system performance. Entropy satisfies the additive property whereby a combination of subsystems will be optimal when their total entropy is minimized, noting that at the maximum value of unity complete disorder is exhibited. Because knowledge pooling is essential in the efficient design of process automation systems, entropy minimization consequently corresponds to knowledge focusing in as few information sources or alternative control actions as practicable. Within Figure 8.12 control actions, C, describe actuation toward processing goals, A, through a progression of processing states, S, that encounter possible disorder events, E.

An ideal control action for subsequent states is defined as that which minimizes the entropy required to achieve a goal state. Figure 8.13 describes how information uncertainty is minimized through online rule acceptance

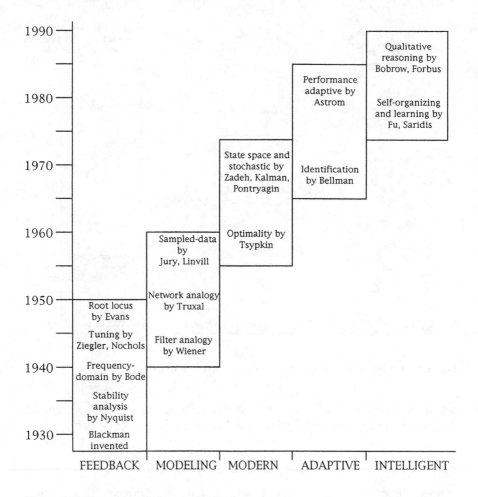

Figure 8.11 Chronology of control advancement.

testing, where no greater than 10% disconfirms are accepted. Anomalous internal and external sources of process disorder often impose unmodeled processing dynamics whose attenuation is influenced by the combined uncertainty of process parameter measurements and *ex situ* planner information uncertainty.

The performance of process apparatus in automation systems may be quantified by measured and modeled device and system errors. That offers useful tolerance accountability by defining the error of closed-loop controlled variables that determine product properties. It follows that this baseline apparatus variability constitutes the irreducible uncertainty of process controlled-variable parameters, illustrated in Figure 8.12, where $\varepsilon_{\text{total}}$ of H_{envir} is expressed as the sum of variances plus mean errors. This expression is derived from system error contributions summarized in the

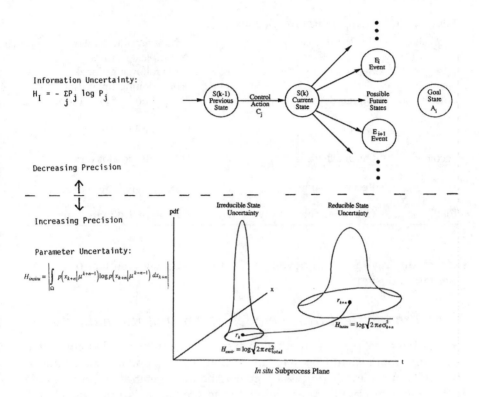

Information Uncertainty:

$$H_I = - \sum_j P_j \log P_j$$

Decreasing Precision

Increasing Precision

Parameter Uncertainty:

$$H_{insitu} = \left| \int_\Omega p\left(x_{k+n} | \mu^{k+n-1}\right) \log p\left(x_{k+n} | \mu^{k+n-1}\right) dx_{k+n} \right|$$

In situ Subprocess Plane

Figure 8.12 Automation information and parameter uncertainty.

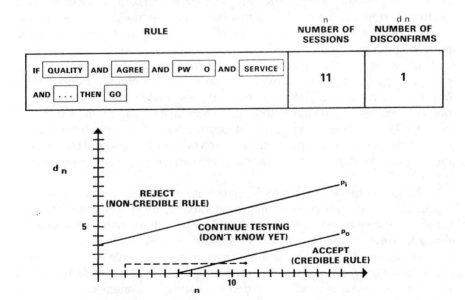

RULE	n NUMBER OF SESSIONS	d n NUMBER OF DISCONFIRMS
IF QUALITY AND AGREE AND PW O AND SERVICE AND ... THEN GO	11	1

Figure 8.13 Expert system rule uncertainty.

Table 8.2 Process Apparatus Error Summary

Acquisition	Analog devices	ε_a	Sensor, amplifier, filter, multiplexer errors
	Signal quality	ε_b	Error following signal conditioning upgrading
Conversion	Digital devices	ε_c	Sample-hold, A/D, computer errors
	Sampling	ε_d	Intersample, aliasing, aperture, sinc errors
Recovery	Output devices	ε_e	D/A, interpolator errors
	Reconstruction	ε_f	Output signal interpolation
Total	Combined uncertainty	ε_{total}	$\sum mean + 1\ \sigma$RSS

accompanying Table 8.2, developed in the preceding chapters on instrumentation design.

8.7 Product features remodeling for quality maturity

Sensor-intensive process automation systems are multivariable information structures that benefit from data attribution beyond fundamental engineering units in achieving processing goals. The interpretation of process data mining to detect product derogation during processing is a priority. Hyperspectral sensing that integrates both spatially and spectrally continuous data is especially useful for product property characterization. Sensor-based feature attribution for product assessment, such as online chemistry and morphology measurements, can be facilitated by pattern recognition to aid product quality growth.

Following process analytical and *in situ* data acquisition, introduced in Figure 8.10, a product features facsimile can be obtained by means of hypothesis evaluation. That structure is trained on material prototypes for microstructure identification through irregularity evaluation, shown in Figure 8.14. This system recognizes definitive output patterns from incomplete input patterns, owing to closed-network decision boundaries in the input space, by capturing the probability distributions of training prototypes.

The advent of controllable molecular material processing systems enables advanced manufacturing performance extending from feature measurements through the implementation of product microstructure control. Essential to that achievement is an *ex situ* planner remodeling capability for ideal process state actuations directed by real-time product features. This system is introduced in Figure 8.15. In operation, migrating product property values result in apparatus physical variables guided by feature measurements to asymptotically meet ideal product goals. A version of *ex situ* planner remodeling for product optimization is demonstrated in Chapter 11.

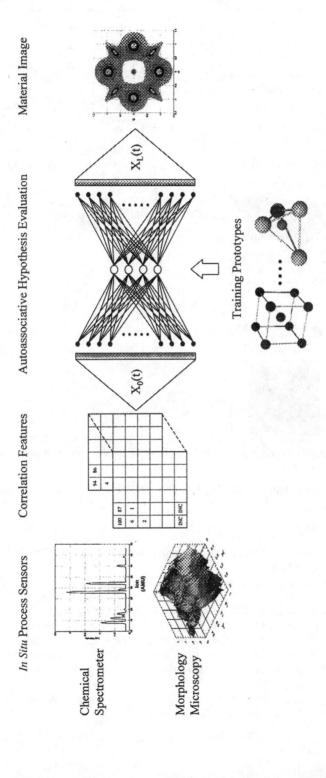

Figure 8.14 Molecular product feature attribution.

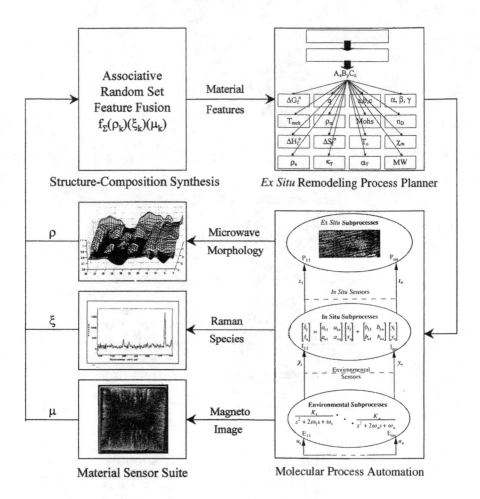

Figure 8.15 Product features *ex situ* planner remodeling.

Bibliography

1. Bobrow, D.G., Ed., *Qualitative Reasoning About Physical Systems*, MIT Press, Cambridge, MA, 1985.
2. Busbee, J., Laube, S.J.P., and Jackson, A.G., "Sensor Principles and Methods for Measuring Physical Properties," *Journal of Materials*, 48(9), 1996 p. 16.
3. Garrett, P.H., *Multisensor Instrumentation 6σ Design*, Wiley, New York, 2002.
4. Garrett, P.H., Jones, J.G., and LeClair, S.R., "Self-Directed Processing of Materials," *Journal of Engineering Applications of Artificial Intelligence*, 12(4), 1999 p. 479.
5. Garrett, P.H., Jones, J.G., Moore, D.C., and Malas, J.C., "Emerging Methods for the Intelligent Processing of Materials," *Journal of Materials Engineering and Performance*, 2(5), 1993 p. 727.
6. Hahn, G.J. et al., "The Impact of Six-Sigma Improvement — A Glimpse Into the Future of Statistics," *American Statistician*, 53(3), 1999 p. 208.

7. Jackson, A.G. and Benedict, M., "Thin Film Growth Simulation using Cellular Automata State Space and Neural Net Methods," USAF Technical Report ML-WP-TR-1999-4028, October 1998.
8. LeClair, S.R., Ed., "Electronic Prototyping Initiative Review," Materials Directorate, Wright-Patterson Air Force Base, Dayton, OH, 1995, 1996, 1997, 1998, 1999, 2000, 2001.
9. LeClair, S.R. and Pao, Y.H., Eds., "Artificial Intelligence in Real-Time Control," *Engr. Applic. Artif. Intell.* October 1998 p. 583, vol. 11, no. 5.
10. Mesarovic, M.D., *Views on General Systems Theory,* Wiley, New York, 1964.
11. Matejka, R.F., "Qualitative Process Automation Language," Aerospace Applications of Artificial Intelligence Conference, Dayton, OH, October 1989.
12. Park, J. and Woods, D., "Discovery Systems for Manufacturing," USAF Technical Report, WL-TR-94-4008, January 1994 Wright Patterson AFB, OH.
13. Saridis, G.N. and Valavanis, K.P., "Analytical Design of Intelligent Machines," *Automatica,* 24(2), 1988 p. 123.
14. Suh, N.P., *Axiomatic Design,* Oxford University Press, Oxford, 2001.
15. Taguchi, G., *Introduction to Quality Engineering: Designing Quality into Products and Processes,* Asian Productivity Organization, Tokyo, 1986.
16. Taylor, W.A., *Optimization and Variation Reduction in Quality,* McGraw-Hill, New York, 1991.

chapter nine

Molecular beam epitaxy semiconductor processing

9.0 Introduction

Improvement in the stability of molecular beam epitaxy (MBE) semiconductor processing is described employing concurrent engineering methods. These improvements support progress in electro-optical devices grown in epitaxial thin-film forms, such as millimeter wave monolithic integrated circuits (MIMIC), and enhances semiconductor manufacturing processes overall. Reduced-variability MBE process control performance is obtained by decoupling the principal sources of effusion cell variability with: (1) identification of trapezoidal proportional-integral-derivative (PID) cell temperature controller tuning values to replace typical quarter-decay tuning values; (2) feedforward-model compensation for cancellation of cell enthalpy transients at shutter openings; and (3) comprehensive measurement and control apparatus error evaluation with identification of each contributing system component. These advances enable the use of spectroscopic ellipsometry for nondestructive *in situ* measurement of surface growth properties and regulation of film thickness.

9.1 Molecular beam epitaxy concurrent engineering

Molecular beam epitaxy has achieved only restricted acceptance as a manufacturing process because it lacks a production *ex situ* planner. However, MBE effusion cell decoupling and control compensation refinements are introduced for improved growth stability and the asymptotic realization of near-defect-free material. The following sections accordingly document the evolutionary contributions to MBE process integration and growth control by means of concurrent engineering.

The limited number of MBE controlled variables—basically effusion cell parameters and exposure times—is a fundamental problem compounded by growth variability issues attributed to processing complexity. This chapter

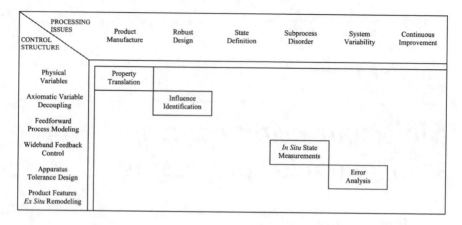

Figure 9.1 Molecular beam epitaxy concurrent engineering.

nevertheless incorporates four of the concurrent engineering linkages introduced in the taxonomy of Figure 8.1, including material property modeling, axiomatic parameter decoupling, *in situ* ellipsometry, and apparatus error analysis, as depicted in Figure 9.1. Systematic improvements for decoupling growth variability from sources of process disorder are described in the robust control design section. A comprehensive MBE effusion cell variability characterization is further provided by a closed-loop cell measurement and control error analysis. That result reconciles combined multiple-cell flux variability on the order of 1%FS 1σ, which verifies adequate flux synchrony to produce lattice-matched film growths.

A next-generation addition to MBE control systems will be an *ex situ* process planner for coordinating recipe execution, cell temperature controller tracking, and *in situ* control reference synthesis. The provision of an *ex situ* planner structure will integrate material characterization data for growth performance and reduce the skilled manpower required for manual MBE operation. Figure 9.2 describes a modular representation for the process and control structure of an MBE machine.

9.2   *MBE material property modeling*

MBE can produce materials from the order of a single atomic layer of deposition to sharp interlayer interfaces required for efficient optoelectronic GaAlAs multiquantum-well lasers. A typical growth rate is 1 monolayer per second, or 1 μm/h. Many practical devices require layers 2-to-8 μm thick, which, accordingly, translates into 2- to 8-h growth times. With example III-V semiconductor crystals, the ratio between atom types defines device stoichiometry. For example, considering $Ga_{0.8}Al_{0.2}As$, every ten atoms of As requires eight of Ga and two of Al. If this stoichiometry is changed, then the respective effusion cell temperatures must be changed correspondingly to alter flux beam equivalent pressures, F. A specific MBE material growth recipe is described in Figure 9.3.

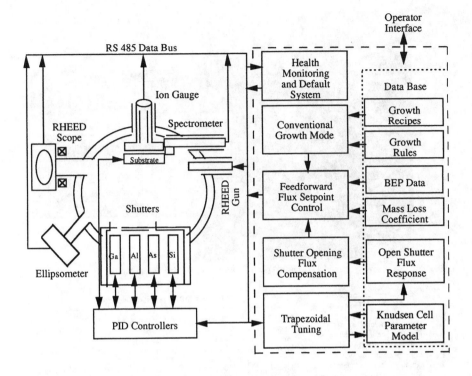

Figure 9.2 Molecular beam epitaxy modular apparatus.

Prior to material growth, MBE machines require calibration to relate Knudsen cell temperature values to required flux, F, at the substrate, while compensating for cell material mass loss with consumption, to resolve a beam equivalent pressure (BEP) of F indicated by ion gauge measurements. During calibration an ion gauge is positioned with a direct view of the cells, but for material production the gauge must be positioned to permit unobstructed substrate growth. Ultralinear flux measurements for control of growth parameters continue to be an ongoing quest.

9.3   Robust process design and PID tuning

Of interest is achieving automatic control of the MBE laboratory process with robustness equivalent to an industrial control system. It is consequently axiomatic to seek MBE machine enhancement involving systematic decoupling of environmental cell flux variability from *in situ* flux subprocess disorder. The described improvements offer greater flux stability plus accuracy of cell temperature selection, and ultimately layer thickness growth repeatability, than original equipment manufacturers typically provide. The preceding Figure 9.2 shows a conventional chamber

Molecular Beam Epitaxy Chamber/InfoScribe Data Visualization

$$F = \frac{A\cos\theta \; P}{\pi r^2 \sqrt{2\pi m k T}}$$

EXAMPLE MBE MATERIAL GROWTH

F	Flux atoms/cm² at wafer	layer 7	500 layers	$Ga_{0.9}Al_{0.1}As$	
A	Knudsen cell aperture area	layer 6	300 layers	GaAs	10^{17} Si doped
θ	Cell-wafer offset angle	layer 5	200 layers	$Ga_{0.8}Al_{0.2}As$	
P	Flux pressure at wafer	layer 4	300 layers	GaAs	10^{17} Si doped
T	Knudsen cell temperature	layer 3	200 layers	$Ga_{0.8}Al_{0.2}As$	
r	Shortest cell-wafer distance	layer 2	300 layers	GaAs	10^{17} Si doped
k	Boltzman's constant	layer 1	1000 layers	$Ga_{0.8}Al_{0.2}As$	
m	Mass of effluent				

Figure 9.3 Molecular beam epitaxy material properties.

process, often maintained at vacuum between 10^{-9} and 10^{-11} Torr during growth, and its associated control system. This is complemented by Figure 9.4 with a hierarchical perspective that shows the interrelation between ascending environmental, *in situ,* and *ex situ* subprocess variables that are mathematically indicated by the decoupled mapping matrices of equation (9.1).

$$\begin{bmatrix} \delta \\ \mu \end{bmatrix} = \begin{bmatrix} P_{11} & 0 \\ 0 & P_{22} \end{bmatrix} \begin{bmatrix} I_{11} & 0 \\ I_{21} & I_{22} \end{bmatrix} \begin{bmatrix} E_{11} & 0 \\ 0 & E_{22} \end{bmatrix} \begin{bmatrix} u_1 \\ u_2 \end{bmatrix} \quad \text{MBE subprocess mapping} \quad (9.1)$$

The effusion cell evaporant elements each have a finite thermal mass such that their respective fluxes respond to both heater excitation temperature input changes with time constant τ_o, nominally 1 sec, and an evaporant load output time constant at shutter opening of τ_2, typically 20 sec. The uncompensated flux transient at shutter opening can appreciably disturb

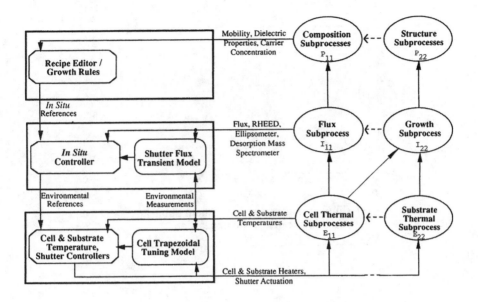

Figure 9.4 Molecular beam epitaxy processing influences.

interlayer growth interfaces. Figure 9.5 accordingly describes a feedforward temperature setpoint compensator algorithm that provides a cell tempera-ture increase prior to shutter opening to cancel this flux transient. Improved flux transient decoupling from 10 to 1% of beam flux values is routinely achieved with this compensator.

The variability of effusion cell flux values during material layer growth is further influenced by the performance of cell temperature control loops. Experimental observation has shown flux repeatability improvement from retuning environmental controller PID values when temperature setpoint changes exceed ±50°C. Optimum PID selection concurrently must provide: (1) control responsiveness with unconditional stability; (2) minimum steady-state error between temperature setpoints and controlled variables; and (3) enhanced transient response following cell process load changes. These three functions are determined, respectively, by proportional band, *P*, integral time, *I*, and derivative time, *D*. Comparison of Figures 9.6 and 9.7 demonstrates an approximate order-of-magnitude performance improve-ment for trapezoidal tuning, as evaluated by the integral-squared error of equation (9.2). This result is attributed to a reduction in coupling between PID controller terms for the trapezoidal PID algorithm and emphasizes the value of precise controller compensation.

$$\varepsilon_{PID} = \frac{1}{t_f - t_o} \int_{t_o}^{t_f} \left(\frac{R_k - T}{\Delta R_k} \right)^2 \bullet dt \bullet 100\% \qquad \text{controller tuning error} \qquad (9.2)$$

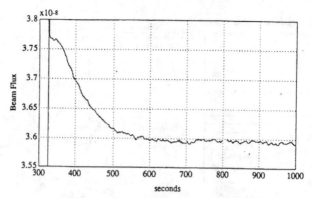

a) Flux at shutter opening without Shutter
Opening Transient Compensation

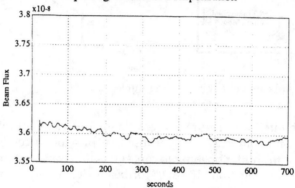

b) Flux at shutter opening with Shutter
Opening Transient Compensation

$$+\Delta R_k = \frac{+\Delta R_k' \cdot \Delta \text{Flux}}{0.9\,\Delta \text{Flux}'} \cdot \left[1 + \left(\frac{\tau_2' - \tau_2}{\tau_2}\right)\right]$$
$$\cdot \exp\left(-t/\tau_2\right)$$

$+\Delta R_k$ Temperature setpoint increment for shutter transient compensation initiated at t_0 minus process lag time.

ΔFlux Shutter opening flux change to steady state.

τ_2 Evaporant mass time constant from ΔFlux.

$+\Delta R_k'$ Arbitrary temperature setpoint increment for open shutter bump test.

$\Delta \text{Flux}'$ Open shutter flux change to steady state for $+\Delta R_k'$.

τ_2' Evaporant mass time constant from $\Delta \text{Flux}'$.

Figure 9.5 Cell enthalpy transient compensation.

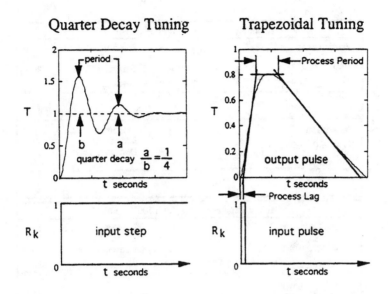

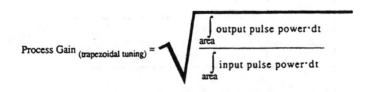

Figure 9.6 Quarter-decay vs. trapezoidal PID tuning.

Comprehensive online process data acquisition by the InfoScribe data archival system, originally developed for MBE machines and shown in Figure 9.3, further reveals flux amplitude oscillations to 8% of average flux values at 10-min intervals. Investigation disclosed that U.S. power utilities execute power generation phase/frequency corrections every 10 min, nominally of ±0.01 Hz. MBE flux oscillations are coupled to this power system correction by two circumstances. One is a common use of nonlinear triac ac

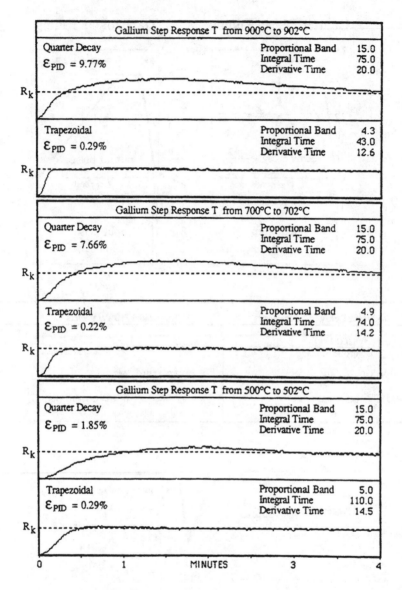

Figure 9.7 Gallium cell temperature control comparison.

power drivers for cell temperature heaters. The second is a flux evaporant cell mass-determined bandwidth, evaluated as $(2\pi \tau_2)^{-1}$ Hz, or 0.008 Hz for typical 20-sec τ_2 values, enabling temperature control loop transmission of these power corrections to cell flux outputs. Decoupling of power frequency corrections for both circumstances is obtained by substitution of 1-kW dc power drivers for triac ac drivers.

9.4 *MBE* in situ *ellipsometry*

A goal of *in situ* sensors is to provide material inspection during processing through nondestructive product evaluation. *In situ* measurement of MBE process mass–momentum–energy quantities is consequently of interest for enabling accurate thin-film growths by means of automatic control. Figures 9.2 and 9.4 describe *in situ* flux and film growth sensors. Ion gauge sensing for cell flux calibration is discussed in Section 9.2. Growth calibration using reflective high-energy electron diffraction (RHEED) can validate growth rates of elemental material layers, for specific cell temperature and exposure times, employing a test substrate illuminated during growth with a scanning electron beam that is sensed by an imaging display. However, RHEED cannot be adapted for online *in situ* control of material growth because of deleterious influences and defects in product quality arising from the electron beam. InfoScribe has also been developed specifically for MBE processes to provide standardized data acquisition amenable to data mining for process improvement purposes. This data format is shown in Figure 9.3.

Multiple-wavelength spectroscopic ellipsometry relies on the physical optics of stratified media for nondestructive *in situ* measurement of surface, film, and substrate growth properties to implement online thickness and composition control. Figure 9.8 illustrates a preferred mechanization, with

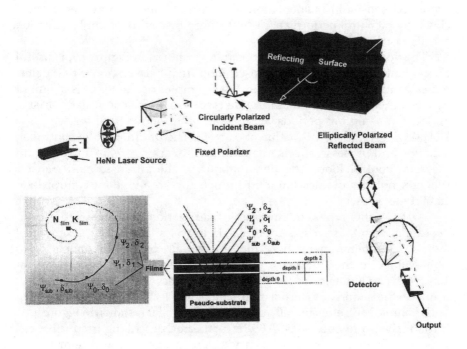

Figure 9.8 In situ ellipsometry for MBE growth control.

freedom from physical film disturbances. In practice, elliptically polarized light reflected from a substrate growth is analyzed for specific electric field amplitude ψ and phase δ pairs in the measurement of material properties. Because the ellipsometry sensor measures ψ and δ values vs. wavelength, substrate optical parameters n, k, and d, respectively, the growth index of refraction, extinction coefficient, and thickness must then be estimated. Inverse solution of these estimates is achieved employing Fresnel equations, while comparing a predicted fit of ψ and δ values to a growth redundantly known by direct measurement for calibration purposes. Online operation is aided by the Levenberg-Marquardt algorithm, with measurement of substrate growth dielectric-function spectra vs. temperature. *In situ* feedback growth control is then implemented by cell temperature adjustment to achieve a goal stoichiometry of interest with layer thickness to ±1 Å accuracy.

9.5 *Instrumentation and control tolerance analysis*

A majority of process control systems include PID controllers for their environmental apparatus to obtain standard control functions including data exchange capabilities. For many applications, controllers acquire process measurements, absent control actuation, because of the utility of their contained sensor instrumentation. Of greater interest, however, is how closed-loop control performance is influenced by controller implementation and its associated process response.

The dominant process time constant, τ_o, shown in Figure 9.9, is useful for evaluating the BW_{CL} −3-dB closed-loop frequency response that determines controlled-variable error. In this example, rise time, t_r, is acquired between the 10% and 90% amplitude response of the controlled variable $c(n)$ in ten sampling periods, with a sampling period T of 0.1 sec ($fs = 10$ Hz), yielding the BW_{CL} value of 0.35 Hz. The product of the controller, actuator, and process gains is lumped into K to approximate unity around the total control loop. The denominator of the z-transformed transfer function notably defines the joint influence of K and T on control loop stability as shown.

A 0 to 1800°C Type-C thermocouple linearization and cold-junction compensation circuit, as well as input 22-Hz lowpass filter, multiplexer signal loss, and sampling attenuation sinc function all contribute nominal mean error values to the Eurotherm digital PID controller error summary of Table 9.1. The influence of BW_{CL} on cell dynamics following excitation, U, results in the dominant interpolation error of 0.174%FS in the error summary. This example MBE effusion cell temperature controller is shown in Figure 9.10. Error analysis provides a 0.45%FS 1σ total error, describing irreducible cell variability, equivalent to 8-bit binary accuracy, based on cell temperature control loop device and system error contributions, which determine the baseline cell performance capability.

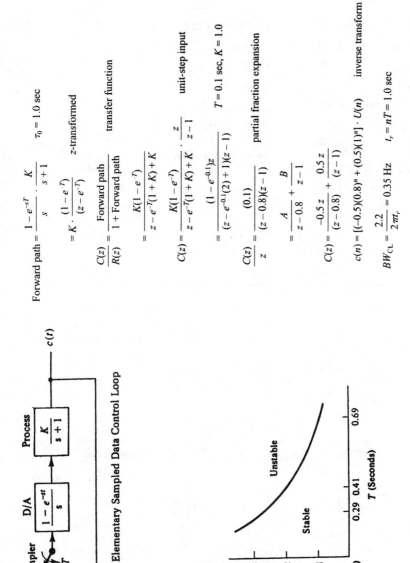

$$\text{Forward path} = \frac{1 - e^{-sT}}{s} \cdot \frac{K}{s+1} \qquad \tau_0 = 1.0 \text{ sec}$$

$$= K \cdot \frac{(1 - e^{-T})}{(z - e^{-T})} \qquad \text{z-transformed}$$

$$\frac{C(z)}{R(z)} = \frac{\text{Forward path}}{1 + \text{Forward path}} \qquad \text{transfer function}$$

$$= \frac{K(1 - e^{-T})}{z - e^{-T}(1 + K) + K}$$

$$C(z) = \frac{K(1 - e^{-T})}{z - e^{-T}(1 + K) + K} \cdot \frac{z}{z - 1} \qquad \text{unit-step input}$$

$$= \frac{(1 - e^{-0.1})z}{(z - e^{-0.1}(2) + 1)(z - 1)} \qquad T = 0.1 \text{ sec}, K = 1.0$$

$$\frac{C(z)}{z} = \frac{(0.1)}{(z - 0.8)(z - 1)} \qquad \text{partial fraction expansion}$$

$$= \frac{A}{z - 0.8} + \frac{B}{z - 1}$$

$$C(z) = \frac{-0.5\,z}{(z - 0.8)} + \frac{0.5\,z}{(z - 1)}$$

$$c(n) = [(-0.5)(0.8)^n + (0.5)(1)^n] \cdot U(n) \qquad \text{inverse transform}$$

$$BW_{\text{CL}} = \frac{2.2}{2\pi t_r} = 0.35 \text{ Hz} \qquad t_r = nT = 1.0 \text{ sec}$$

Figure 9.9 Controller closed-loop bandwidth evaluation.

Table 9.1 Control Instrumentation Error Summary

Element	$\varepsilon_{\%FS}$	Comment
Sensor	$\overline{0.011}$	Linearized thermocouple
Interface	$\overline{0.032}$	Cold-junction compensation
Amplifier	0.103	OP-07A
Filter	$\overline{0.100}$	Table 3.5
Signal quality	0.009	60 Hz ε_{coh}
Multiplexer	$\overline{0.011}$	Average transfer error
A/D	0.020	14-bit successive approximation
D/A	0.016	14-bit R-2R converter
Noise aliasing	0.000049	–85-dB AMUX crosstalk from 40 mV @ 20 kHz
Sinc	$\overline{0.100}$	Average attenuation over BW_{CL}
Intersample	0.174	Interpolated by BW_{CL} from process τ_o
	$\overline{0.254}$%FS	$\Sigma mean$
ε_C	0.204%FS	1σ RSS
	$\overline{0.458}$%FS	$\Sigma \overline{mean} + 1\sigma RSS$

Sensor

Type-C thermocouple 17.2 µV/°C postconditioning linearization software

$$\frac{\overline{0.2°C}}{1800°C} \bullet 100\% = \overline{0.011}\%FS$$

Interface

AD 590 temperature sensor cold-junction compensation

$$\frac{\overline{0.5°C}}{1800°C} \bullet 100\% = \overline{0.032}\%FS$$

Signal Quality

$$\varepsilon_{coh} = \frac{V_{cm}}{V_{diff}} \bullet \left[\frac{R_{diff}}{R_{cm}}\right]^{1/2} \bullet \frac{A_{V_{cm}}}{A_{V_{diff}}} \bullet \left[1+\left(\frac{f_{coh}}{f_c}\right)^{2n}\right]^{-1/2} \bullet 100\%$$

$$= \frac{(1\ V_{rms}2\sqrt{2})_{pp}}{31\ mV_{dc}} \bullet \left[\frac{80\ M\Omega}{200\ G\Omega}\right]^{1/2} \bullet \frac{0.02}{132} \bullet \left[1+\left(\frac{60\ Hz}{22\ Hz}\right)^2\right]^{-1/2} \bullet 100\%$$

$$= 0.009\%FS$$

$\varepsilon_{ampl_{RT1}}$	**OP-07A**
V_{os}	$\overline{10}\ \mu V$
$\dfrac{dV_{os}}{dT} \bullet dT$	$2\ \mu V$
$I_{os} \bullet R_i$	$\overline{3}\ \mu V$
$V_{N_{pp}}$	$4.4\ \mu V$
$f(Av) \bullet \dfrac{V_{o_{FS}}}{Av_{diff}}$	$\overline{3}\ \mu V$
$\dfrac{dAv}{dT} \bullet dT \bullet \dfrac{V_{o_{FS}}}{Av_{diff}}$	$15.5\ \mu V$
$\Sigma\ \overline{mean} + 1\sigma$ RSS	$(\overline{16} + 16)\mu V$
$X\dfrac{Av_{diff}}{V_{o_{FS}}} \bullet 100\%$	0.103%FS

Analog Multiplexer		
Transfer error		$\overline{0.01}\%$
Leakage		0.001
Crosstalk		0.00005
ε_{AMUX}	$\Sigma\ \overline{mean} + 1\sigma$RSS	$0.011\ \%$FS

14-Bit A/D		
Mean integral nonlinearity (1 LSB)		$\overline{0.006}\%$
Noise + distortion (–80 dB)		0.010
Quantizing uncertainty ($\frac{1}{2}$ LSB)		0.003
Temperature coefficients ($\frac{1}{2}$ LSB)		0.003
$\varepsilon_{A/D}$	$\Sigma\ \overline{mean} + 1\sigma$RSS	0.020%FS

14-Bit D/A		
Mean integral nonlinearity (1 LSB)		$\overline{0.006}\ \%$
Noise + distortion (–80 dB)		0.010
Temperature coefficients ($\frac{1}{2}$ LSB)		0.003
$\varepsilon_{D/A}$	$\Sigma\ \overline{mean} + 1\sigma$RSS	0.016%FS

Noise Aliasing

$$\varepsilon_{\text{coherent alias}} = \text{Interference} \bullet \text{AMUX crosstalk} \bullet \text{sinc} \bullet 100\%$$

$$= \frac{V_{coh}}{V_{o_{FS}}} \bullet -85 \text{ dB} \bullet \left(\frac{mf_s - f_{coh}}{f_s}\right) \bullet 100\% \quad m \text{ defined at } f_{coh}$$

$$= \frac{40 \text{ mV}}{4096 \text{ mV}} \bullet (0.00005) \bullet \text{sinc}\left(\frac{2000 \bullet 10 \text{ Hz} - 20 \text{ kHz}}{10 \text{ Hz}}\right) \bullet 100\%$$

$$= 0.000049\%\text{FS}$$

Sinc

$$\varepsilon_{\text{sinc}} = \frac{1}{2}\left(1 - \frac{\sin \pi \, \text{BW}_{\text{CL}}/f_s}{\pi \, \text{BW}_{\text{CL}}/f_s}\right) \bullet 100\%$$

$$= \frac{1}{2}\left(1 - \frac{\sin \pi \, 0.35/10}{\pi \, 0.35/10}\right) \bullet 100\%$$

$$= \overline{0.100} \, \%\text{FS}$$

Controlled Variable Interpolation

$$\varepsilon_{\Delta V} = \left[\frac{V_{O_{FS}}^2}{V_S^2 \bullet \left\{\text{sinc}^2\left(1 - \frac{\text{BW}_{\text{CL}}}{f_s}\right) \bullet \left[1 + \left(\frac{f_s - \text{BW}_{\text{CL}}}{\text{BW}_{\text{CL}}}\right)^2\right]^{-1} + \text{sinc}^2\left(1 + \frac{\text{BW}_{\text{CL}}}{f_s}\right) \bullet \left[1 + \left(\frac{f_s + \text{BW}_{\text{CL}}}{\text{BW}_{\text{CL}}}\right)^2\right]^{-1}\right\}}\right]^{-1/2} \bullet 100\%$$

$$= \left[\frac{4.096V^2}{(4.096V)^2 \bullet \left\{\left[\frac{\sin \pi\left(1 - \frac{0.35 \text{ Hz}}{10 \text{ Hz}}\right)}{\pi\left(1 - \frac{0.35 \text{ Hz}}{10 \text{ Hz}}\right)}\right]^2 \bullet \left[1 + \left(\frac{10 \text{ Hz} - 0.35 \text{ Hz}}{0.35 \text{ Hz}}\right)^2\right]^{-1} + \left[\frac{\sin \pi\left(1 + \frac{0.35 \text{ Hz}}{10 \text{ Hz}}\right)}{\pi\left(1 + \frac{0.35 \text{ Hz}}{10 \text{ Hz}}\right)}\right]^2 \bullet \left[1 + \left(\frac{10 \text{ Hz} + 0.35 \text{ Hz}}{0.35 \text{ Hz}}\right)^2\right]^{-1}\right\}}\right]^{-1/2} \bullet 100\%$$

$$= \left[\frac{1}{\left(\frac{0.110}{3.03}\right)^2 \bullet (0.001313) + \left(\frac{-0.1094}{3.251}\right)^2 \bullet (0.001142)}\right]^{-1/2} \bullet 100\%$$

$$= 0.174\%\text{FS}$$

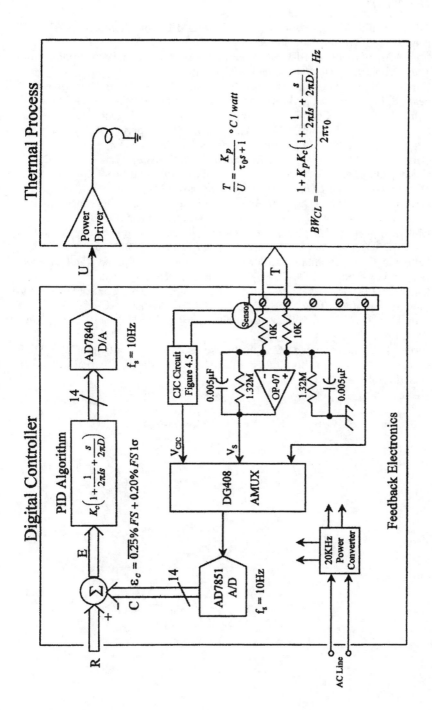

Figure 9.10 Cell temperature controller.

Bibliography

1. Adams, S. and Heyob, J., "Info Scribe," WL/MLIM Technical Reports ASC95-1525, August 1995. Wright-Patterson AFB.
2. Cho, A.Y., "Growth of III-V Semiconductors by Molecular Beam Epitaxy and Their Properties," *Thin Solid Films*, Vol. 100, 1983.
3. Currie, K.R. and LeClair, S.R., "Self-Improving Process Control for Molecular Beam Epitaxy," *Journal of Advances in Manufacturing Technology*, 8, 244, 1993.
4. Eyink, K.G. et al., "Combined Use of Computational Intelligence and Materials Data for Online Monitoring and Control of MBE Experiments," *Engineering Applications of Artificial Intelligence*, no. 5, vol 11, 1998 p. 587.
5. Garrett, P.H., *Multisensor Instrumentation 6σ Design*, Wiley, New York, 2002.
6. Garrett, P.H. et al., "Decoupled Flux Control for Molecular Beam Epitaxy," *IEEE Transactions on Semiconductor Manufacturing*, 6(4), 1993 p. 348.
7. Heyob, J.J., *The Process Discovery Autotuner*, Master's Thesis, Department of Electrical and Computer Engineering, University of Cincinnati, OH, June 1991.
8. Ishikawa, T. et al., "Application of Modern Control to Temperature Control of the MBE Systems," *Japanese Journal of Applied Physics*, 29(3), 1990.
9. Patterson, O.D., "Qualitative Control of Molecular Beam Epitaxy," Self Directed Control Workshop, WRDC-TR-90-4123, Dayton, OH, May 1990.
10. Patterson, O.D. et al., "Progress Toward a Comprehensive Control System for Molecular Beam Epitaxy," USAF WL-TR-92-4091, August 1992.
11. Tompkins, H.G. et al., *Spectroscopic Ellipsometry and Reflectometry*, Wiley, New York, 1999.
12. Vlcek, J.C. and Fonstad, C.G., "Precise Computer Control of the MBE Process — Application to Graded InGaAlAs/InP Alloys," *Journal of Crystal Growth*, Vol. 111, 1991 p. 55.

chapter ten

Aerospace composites rule-based manufacturing

10.0 Introduction

A qualitative process automation system for the autoclave cure of epoxy aerospace composites is described, including contributions from material science, computational intelligence, and automatic control. This system senses internal product properties online during processing and employs a rule-based *ex situ* planner programmed by cure experts' domain knowledge describing ideal process curing. *In situ* process sensor fusion information accordingly is interpreted for cure state determination with feedback execution to attain product goals while attenuating unmodeled processing dynamics. Concurrent engineering provides a framework that captures the essential implementation methods utilized in this application example, wherein A-10 Thunderbolt II aircraft leading edge structures, shown in Figure 10.1, are successfully manufactured. Demonstrated is a product-directed qualitative process automation system that includes the elements for self-improvement through an autonomy to execute nonprescriptive control-path changes during processing.

10.1 Composite manufacturing concurrent engineering

The concurrent engineering linkages shown in Figure 10.2 represent the qualitative process automation (QPA) system described for aerospace composites manufacturing. The implementation of these linkages maps processing issues to control structures to enable process performance and product values not achievable in their absence. Autoclave curing of thermosetting epoxy composites for aerospace components is the initial application of QPA, which is embodied as a rule-based computationally intelligent mechanization.

This system incorporates four concurrent engineering linkages, including physical property modeling, a rule-based *ex situ* planner, *in situ* process state control, and apparatus tolerance analysis. Because composite cure processing

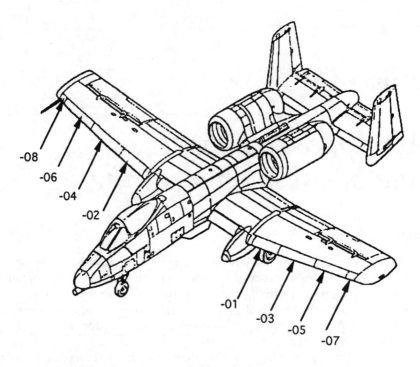

Figure 10.1 A-10 aircraft composite leading edges.

is constrained to two principal controlled variables, temperature and pressure, axiomatic parameter decoupling is not considered. Although residual voids also continue to be problematic in some composite cure applications, which emphasizes the need for continuous improvement in quality, that is not a problem with this application and therefore not included.

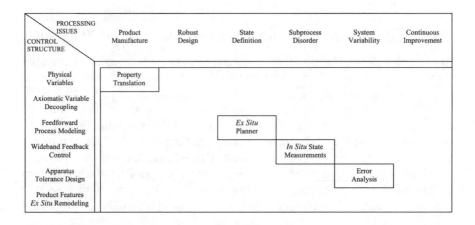

Figure 10.2 Composite cure concurrent engineering.

10.2 Composite cure properties and apparatus

Modeling of epoxy composite structural materials is product-performance oriented in terms of fiber/resin specifications, layup orientation, precure compaction, postcure product physical properties such as tension strength, and residual laminate voids. Interactions among process variables are illustrated by the diagram of Figure 10.3, revealing that only autoclave temperature and

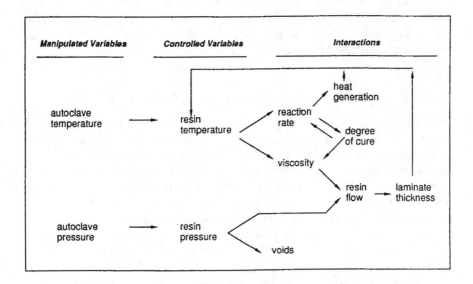

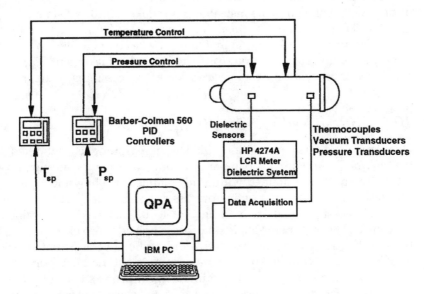

Figure 10.3 Composite cure variables and apparatus.

pressure are available as controlled variables and their gradients internal to laminates are the principal influence on curing uniformity.

A detailed definition for process control execution is described by an expert rule-based interpretation of essential product properties. This is expressed in terms of corresponding process actuator values by processing sequences that seek product composition and microstructure requirements while accommodating process disorder events. Attained product properties are reconciled with actuator values employing sensed process states, relative to ideal process states described in the rule-based *ex situ* planner section. Thus, QPA responds to sensed processing events by achieving or preventing states online as required to obtain product property goals, employing nonprescriptive control-path changes as required.

10.3 Computational qualitative reasoning

Qualitative reasoning replaces quantitative variables and algorithms, often represented as differential equations, by a combination of symbolic and quantitative equivalences for predicting physical processes. For control applications, process planners provide a state graph describing system behavior employing a qualitative calculus that implements four tasks: determining system states, arbitrating appropriate state influence actions, resolving conflicts, and translating quantitative-to-symbolic sensor and actuator values. This schema was developed to overcome the limitations of traditional prescriptive process control systems, which lack capabilities for attenuating unmodeled process disorder.

An intelligent rule-based model therefore enables control as a function of process events and goals instead of an inflexible process state time progression. That capability is crucial for unexpected changes detected within a material during processing so that they may be compensated by changing the state control path, ensuring processing goal achievement. Figure 10.4 shows the framework for this object-oriented processing scheduler.

10.4 Rule-based ex situ planner

Qualitative process automation rule-based reasoning is embodied as a feedforward *ex situ* process planner providing ideal composite cure process states. The objects contained within this planner are shown in Figure 10.5. Plans contain *episodes* that define process variables over a cure cycle, each of which describe goal *states* either yet to be achieved, maintained, or prevented. During operation, ideal process episodes are executed sequentially, with each providing respective descriptions of anticipated process events. This expert model of the cure process is alternatively portrayed by the Thinker module shown in the hierarchical process control influences diagram of Figure 10.6 that is interfaced to process apparatus through Blackboard memory resources.

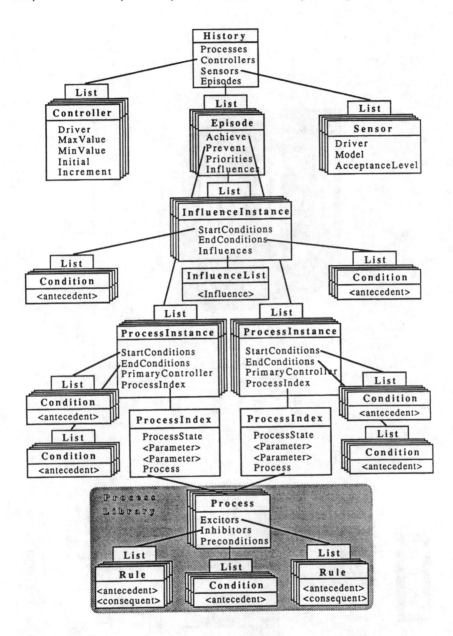

Figure 10.4 Qualitative reasoning *ex situ* model.

The main Cure Plan has three episodes: namely precure Resin Flow, Begin Gel, and End Cure. *In situ* state feedback control is initially activated under the Resin Flow episode, employing embedded laminate thermocouple temperature and dielectric resin viscosity sensors to enable uniform gradient control. At Cure Plan initiation, autoclave temperature is

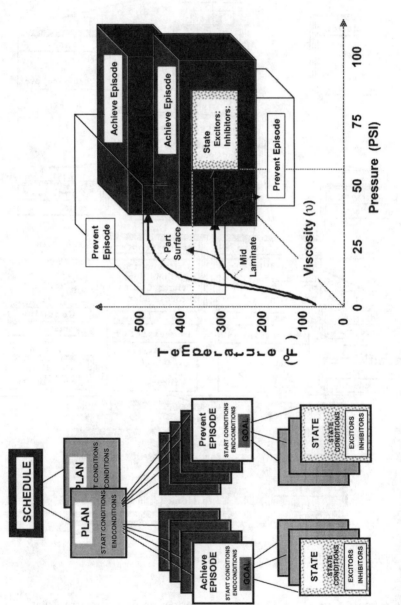

Figure 10.5 Ex situ planner rule-directed control.

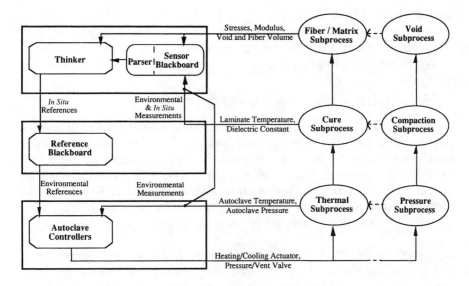

Figure 10.6 Composite cure processing influences.

increased in 10°F increments to 250°F every 30 sec, until laminate viscous flow is inferred from a sharp decrease in measured dielectric impedance from megohms to kilohms, also denoting an increase in resin ionic conductivity. That event prompts an exit from the Resin Flow episode, depicted by the sensor fusion cure data of Figure 10.7.

The Begin Gel episode is then activated, with an increase in laminate temperature to 350°F to realize a solid matrix for transferring loads between fibers, and progresses with autoclave pressure application of typically 85 psig (pounds per square inch gauge) at constant elevated temperature to

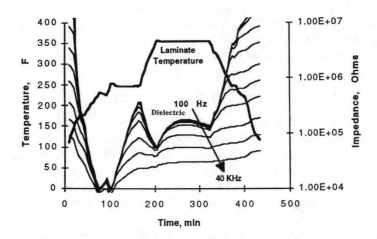

Figure 10.7 In situ sensor fusion cure data.

obtain laminate void collapse and chemical bond crosslinking. End Cure episode initialization invokes a Cooldown Plan, which decreases autoclave temperature to ambient in 10°F increments every 30 sec accompanied by an autoclave pressure dump.

10.5   In situ *sensor fusion cure control*

Process automation systems are multivariable information structures that often require data attribution beyond apparatus environmental parameters to achieve processing goals. For example, sensor measurements representing material mass-momentum-energy parameters are essential for materials property characterization. Further, the need for attenuation of both internal and external sources of processing disorder extends the necessity for sensor measurements beyond material characterization to real-time *in situ* state control.

Embedded thermocouple arrays for sensing laminate temperature gradients offer more accurate quantitative measurements of product cure uniformity, and therefore laminate strength, than is possible with temperatures acquired at process apparatus boundaries. Alternative sensor fusion measurement of product cure employs laminate imbedded dielectric impedance sensors to aid evaluation of product cure progression. Composite cure *in situ* control is further enhanced because the laminate process reaction rate is well within the environmental *in situ* subprocess control bandwidth.

Although some product features cannot be observed directly, they may be inferred from process inputs, outputs, and the process planner model. Crucial determination of the end of the Begin Gel episode and initiation of the End Cure episode is signified by a decrease in dielectric impedance at low 100-Hz frequencies for a laminate temperature of 350°F or greater, with reference to the embedded sensor data of Figure 10.7. Qualitative control is accordingly structured to describe process goal states, resolution of possible conflicts, and translation of sensor and actuator values into symbolic equivalences, as illustrated in Figure 10.5. This process automation example illustrates the collaborative utility of multisensor data fusion for ambiguity resolution and property retrieval unavailable from single-sensor data.

10.6   *Temperature control tolerance analysis*

A concern for control systems is the instability introduced when closed-loop measurement and control errors exceed the equivalent percentage of full-scale (FS) amplitude actuator value changes. Referring to the temperature instrumentation error summary of Table 10.1, note that the combined error of 0.56%FS is equivalent to 3°C uncertainty, resulting in the selection of temperature actuation values that will execute stably in 10°C increments or greater.

This error summary is derived from 500°C measurement scaling, for a Type-J thermocouple maximum signal of 27.388 mV; times a gain of 366 that

Table 10.1 Temperature Instrumentation Error Summary

Element	$\varepsilon_{\%FS}$	Comment
Sensor	0.01	Linearized Type-J thermocouple
Interface	0.10	Cold junction compensation
Amplifier	0.08	Analog Devices 3B47
Filter	0.10	Table 3-5
Multiplexer	0.01	Transfer error
Sample hold	0.02	1 μsec acquisition time
A/D	0.08	12-bit successive approximation
Sinc	0.01	Sampling attenuation
D/A	0.05	12-bit converter
Intersample	0.19	$f_s = 66$ mHz with 1-pole interpolation
Actuator interpolator	0.10	Control element nonlinearity
ε_{total}	0.56%	$\sum mean + 1 \ \sigma$RSS

equals 10 V full-scale output corresponding to 100%. The 0.56%FS error accordingly represents the temperature-controlled variable irreducible state uncertainty, and that variability cannot be further reduced through controller compensation tuning methods. Models upon which this error summary are based are detailed in the preceding chapters on instrumentation design.

Bibliography

1. Abrams, F.L., Lagnese, T.L., LeClair, S.R., and Park, J.B., "Qualitative Process Automation for Autoclave Curing of Composites," USAF WL-TR-87-4083, 1987 Wright-Patterson AFB, OH.
2. Beadles, J.R. and Spellman, G.P., "Sensing Technology for Pressure Flow, Viscosity and Moisture Content Monitoring in Autoclave Environments," Lawrence Livermore National Laboratory Report, UCRL-ID-110108, Livermore, CA, 1992.
3. Brand, R.G., Brown, P.W., and McKague, E., "Processing Science of Epoxy Resin Composites," Air Force Materials Laboratory Report, AFWL-TR-83-4124, Dayton, OH, 1983 Wright-Patterson AFB, OH.
4. Carpenter, J.F., "Viscosity Behavior of Composite Resins," Naval Air Development Center Report, NADC-86083-60, Warminster, PA, 1986.
5. Ciriscioli, P.R. and Springer, G.S., "Dielectric Cure Monitoring—A Critical Review," *SAMPE Journal*, Vol. 25, p. 35, 1989.
6. Forbus, K., "Qualitative Process Theory," *Qualitative Reasoning About Physical Systems*, Bobrow, D.G., Ed., MIT Press, Cambridge, MA, 1985, p. 85.
7. Garrett, P., Lee, C.W., and LeClair, S.R., "Qualitative Process Automation vs. Quantitative Process Control," American Control Conference Proceedings, Minneapolis, MN, 1987, p. 38.
8. Hunston, D. et al., "Assessment of the State-of-the-Art for Process Monitoring Sensors for Polymer Composites," National Institute of Standards and Technology Report, NISTIR 4514, 1991.

9. Kays, A.O., "Exploratory Development on Processing Science of Thick-Section Composites," Air Force Materials Laboratory Report, AFWL-TR-85-4090, Dayton, OH, 1985.

10. Lagnese, T. and Matejka, R., "A Representational Language for Qualitative Process Control," First International Conference on Industrial & Engineering Applications of Artificial Intelligence, IEA/AIE, Tullahoma, TN, June 1988.

11. LeClair, S.R., Abrams, F.L., and Matejka, R.F., "Qualitative Process Automation: Self-Directed Manufacture of Composite Materials," *AI EDAM*, 3(2), 125, 1989.

12. Matejka, R.F., A Programming Environment for Qualitative Process Control, Master's Thesis, Electrical and Computer Engineering, University of Cincinnati, OH, June 1988.

13. Park, J., "Toward the Development of a Real-Time Expert System," Rochester FORTH Conference Proceedings, Rochester, NY, 1986, pp. 23–33.

14. Warnock, R.B., "Application of Self-Directed Control to the Curing of Advanced Composites," USAF Advanced Composites Program Office, SM-ALC/TIEC McClellan AFB, CA, 1991.

15. Warnock, R.B. and LeClair, S.R., "Qualitative Process Automation for Autoclave Curing of Composites," USAF WL-TR-92-4085, August 1992.

chapter eleven

Fuzzy logic laser deposition superconductor production

11.0 Introduction

Laser-ablated material for stoichiometric thin-film deposition enables the manufacture of engineered crystalline products ranging from superconductors to solid lubricants. Acquired process data of mass–momentum–energy quantities provide online information for the deployment of a real-time *ex situ* planner. This feedforward model directs the physical process over an ideal control trajectory while simultaneously augmenting product quality through process remodeling. Achieving processing goals is aided by stabilizing boundary conditions with defined-accuracy process data measurements, including wideband spectrometer *in situ* fuzzy control of the laser ablation plume. The effectiveness of concurrent engineering design is demonstrated for accommodating the complexity of laser deposition processing, wherein crosscoupled intersubprocess parameter influences are beneficially absent.

11.1 Superconductor processing concurrent engineering

The complexity accommodation of product and process modeling, control technology selection, and data error analysis methods are effective concurrent engineering linkages for laser-deposited material processing. Six linkages are involved in mapping process design issues to control structures, described by Figure 11.1. The material property translation and stoichiometric realization capabilities of this process are especially aided by the robust and inherently decoupled intersubprocess influences described.

Creating an effective feedforward *ex situ* planner for continuous production is difficult for this process owing to the absence of defining laser deposition models. This identification challenge is met by a process data observer employing measurements provided by separate mass–momentum–energy deposition growth microbalance, spectrometer plume density,

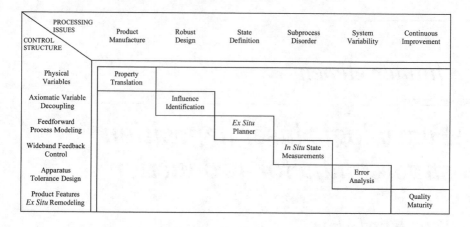

Figure 11.1 Laser-deposited material concurrent engineering.

and incident laser energy sensors. These measurements are then imported into an observer differential equation formulation to generate ideal *in situ* state trajectory predictions.

Achieving laser deposition processing goals is assisted by stabilizing boundary conditions with environmental apparatus regulation and wideband feedback fuzzy control employing emission-spectrometer-sensed plume spectra for compensating process disorder. Fuzzy control benefits from the encoding of expert operator knowledge for maximizing plume output even with nonlinear and only partially measurable process parameters. In fact,

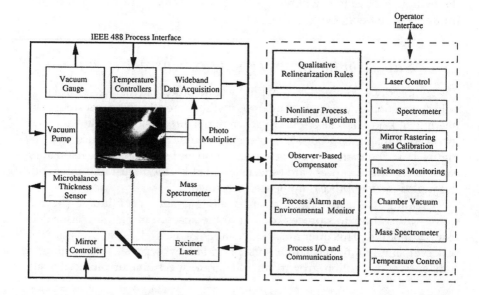

Figure 11.2 Modular laser-deposited material apparatus.

computational intelligence is implemented in this application example not only in the form of fuzzy logic for plume control, but also with separate expert rules that govern *ex situ* planner process remodeling decisions.

Hierarchical process ascent from environmental to *in situ* to *ex situ* subprocess levels demonstrates the principle of decreasing accuracy with increasing complexity, respectively, from 9-bit environmental laser energy to 7-bit *in situ* plume spectrometry to 5-bit *ex situ* product microbalance thickness sensor accuracy evaluations. These sensed values provide essential indices of process performance and product quality to aid enhanced product processing, enabled by *ex situ* planner observer remodeling through online product feature assessment. Offline product quality evaluation is further achieved by Raman spectroscopy using photon scattering understanding with infrared (IR) laser excitation and detection. A modular description of process apparatus is shown by Figure 11.2.

11.2 Material property modeling

Tribological solid lubricant films for spacecraft applications require the realization of precise chemical composition, microstructure, and thickness employing highly energetic atomic bonding that can be effectively met by laser-deposited material processing. Molybdenum disulfide, MoS_2, is a dicalcogenide candidate possessing a crystalline planar microstructure. Its weakly connected Van der Waals force provides a sliding friction plane from –200 to 700°F of 0.01 μ sliding in vacuum and 0.1 μ sliding at atmosphere that is also radiation stable. More complex diamond-like carbon (DLC) ceramic-metal films employ multisource processing, combining laser and magnetron deposition, for example, to produce Ti-TiC-DLC lubricant materials.

Yttrium barium copper oxide (YBCO) is a ceramic superconductor material with a unit cell structure shown in Figure 11.3 having a critical temperature of 92 K. The ideal compound, $YBa_2Cu_3O_{6.85}$, is sensitive to stoichiometric and

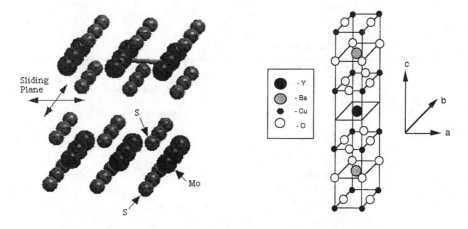

Figure 11.3 Tribology and superconductor materials models.

morphologic variations that benefit from the processing capabilities inherent in laser deposition apparatus directed by hierarchical process control methods for consistent manufacture.

11.3 Robust axiomatic process decoupling

A model of laser deposition processing that characterizes energy transformations at transition boundaries is shown in Figure 11.4. Coherent ultraviolet laser fluence creates energetic electric and kinetic molecular plume species, including chemical exchange reactions, whose resultant thin-film substrate deposition may also extend to recombinant material. Plume density $n(x, y, z, t)$ is modeled by Singh's particle distribution equation (11.1), where τ denotes laser pulse length, N_t the number of particles per pulse, and $X(t)$, $Y(t)$, and $Z(t)$ Cartesian basis functions of material distribution. Processing is typically carried out at 10^{-8} Torr vacuum to ensure purity.

$$n(x,y,z,t) = \frac{N_t t}{\sqrt{2}\pi^{\frac{3}{2}}\tau X(t)Y(t)Z(t)} \exp\left\{-\frac{x^2}{2X(t)^2} - \frac{y^2}{2Y(t)^2} - \frac{z^2}{2Z(t)^2}\right\} \quad (11.1)$$

Material processing complexity increases by the uncertainty encountered in achieving functional requirements, thereby producing growth in information content. Fortunately, processing complexity is axiomatically reducible by means of decoupled system design. This is demonstrated for laser-deposited material by a process structure that meets functional requirements with reduced complexity. Suh's axiomatic design method beneficially provides process parameter minimization in contrast to algorithmic design methods. The multiple regions of laser deposition processing shown by Figure 11.4 accordingly benefit from division into simpler

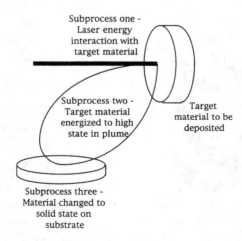

Figure 11.4 Laser deposition subprocess model.

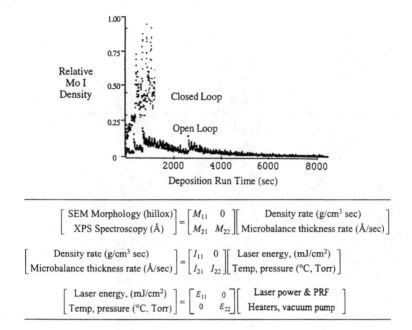

Figure 11.5 Decoupled subprocess parameters.

subprocesses offering more precise parameter definitions. A preferred hierarchical subprocess control system design, detailed in Figure 11.6, is notably absent crosscoupled intersubprocess parameter influences that invoke inefficient process control iteration.

Analysis of this system reveals decoupled parameter influences, definable by their equivalent Gauss reduction to zero off diagonal covariance mapping matrix terms, as revealed in Figure 11.5. These decoupled intersubprocess influences provide improved system linearity and processing effectiveness by achieving greater closed-loop control plume spectrometer density. Parameter decoupling significantly reduces the required iteration of controlled variables for achieving processing goals.

11.4 *Observer remodeled* ex situ *planner*

Ex situ planner structures incorporate feedforward production models that provide event-based control references along ideal process state trajectories. They provide the capability to achieve product goals with greater robustness to processing dynamics than *in situ* or environmental control alone. Environmental regulation allows only sparse representation of the final product, whereas *in situ* control is more comprehensive but limited to segmented realization of final product specifications. *Ex situ* planner representations are more comprehensive but difficult to synthesize for laser-deposited material processing owing to the absence of defining first-principles process models. It is nevertheless possible for mass–momentum–energy process

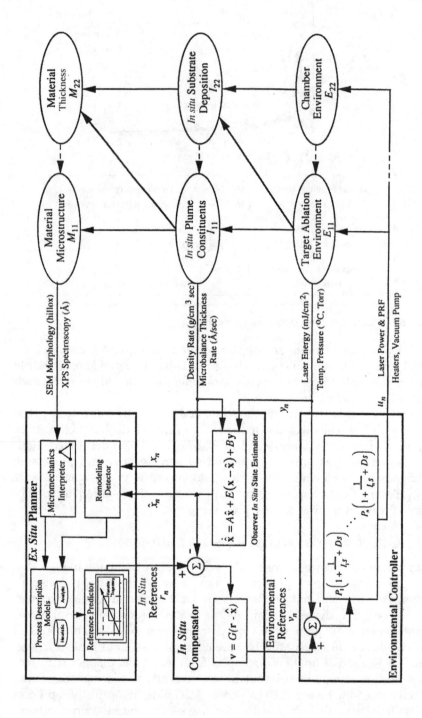

Figure 11.6 Laser-deposited material processing influences.

parameters to be empirically modeled for each of the three primary subprocesses shown in Figure 11.4, as described.

Physical variable sensor measurements for these subprocesses include, respectively, incident target laser energy, plume emission spectrometer species density, and substrate deposition microbalance thickness. Examples of these sensed process data are introduced by Figure 11.5, the defined accuracy of which, derived subsequently, provide indices of system performance. Process nonlinearity is least prevalent for laser target environmental subprocesses, which constitute intersubprocess outputs, y, approximating environmental setpoint reference inputs, v. The *in situ* plume subprocess is more nonlinear and critical to deposition performance, and consequently requires a more comprehensive control implementation to achieve product specifications. This includes careful coordination between the design of the *ex situ* planner and *in situ* compensator.

State-space representations are difficult to characterize for actual processes because of identification challenges, especially without accurate process models. However, an *in situ* subprocess state-space model estimator may be synthesized employing empirical process data to create an observer-based feedback controller. This is described by Figure 11.7, which

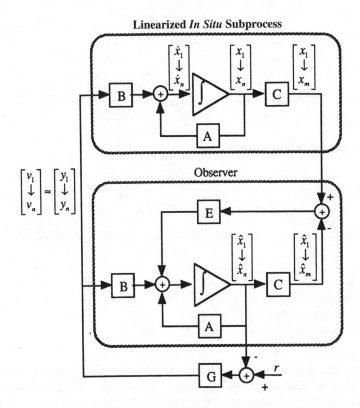

Figure 11.7 *In situ* subprocess observer estimator.

diagrams the *in situ* subprocess observer, where *in situ* subprocess measurements, x, are steered in response to *ex situ* planner *in situ* references, r, and *in situ* compensator matrix, G, as influenced by the observer feedback controller matrix, E, and environmental subprocess outputs, y.

A majority of control system intelligence resides in the observer *in situ* state estimator model. *Ex situ* planner identification is also obtained from this data feature space employing Gaussian basis function state equations, which describe thickness growth rate, $\dot{m}$ (Å), and plume density, $\dot{a}$ (g/cc), that direct environmental control of laser energy, e (mJ/cm^2), and repetition rate, p (Hz). These parameters are imported into a linearizable differential equation matrix formula, equation (11.4), defining *in situ* plume and thickness subprocess states. With process migration, repeated relinearization is exercised to maintain controllability. The decision to relinearize is consequently instituted by computational intelligence rules described next.

Thickness basis function:

$$\dot{m} = \Sigma k_m \exp \left\{ \frac{\left|m - \mu_{mh}\right|^2}{2\sigma_{mh}^2} - \frac{\left|a - \mu_{ah}\right|^2}{2\sigma_{ah}^2} - \frac{\left|e - \mu_{eh}\right|^2}{2\sigma_{eh}^2} - \frac{\left|p - \mu_{ph}\right|^2}{2\sigma_{ph}^2} \right\} \quad (11.2)$$

Plume basis function:

$$\dot{a} = \Sigma k_a \exp \left\{ \frac{\left|m - \mu_{mh}\right|^2}{2\sigma_{mh}^2} - \frac{\left|a - \mu_{ah}\right|^2}{2\sigma_{ah}^2} - \frac{\left|e - \mu_{eh}\right|^2}{2\sigma_{eh}^2} - \frac{\left|p - \mu_{ph}\right|^2}{2\sigma_{ph}^2} \right\} \quad (11.3)$$

Linearized process state:

$$\begin{bmatrix} \dot{m} \\ \dot{a} \end{bmatrix} = \begin{bmatrix} f_1\,(m,a,e,p) \\ f_2\,(m,a,e,p) \end{bmatrix} = \frac{\partial f}{\partial (m,a)}\Bigg|_{\substack{m_o,a_o \\ e_o,p_o}} \begin{bmatrix} \Delta m \\ \Delta a \end{bmatrix} + \frac{\partial f}{\partial (e,p)}\Bigg|_{\substack{m_o,a_o \\ e_o,p_o}} \begin{bmatrix} \Delta e \\ \Delta p \end{bmatrix} \quad (11.4)$$

where
m = microbalance sensed thickness (Å)
a = spectrometer sensed plume density (g/cc)
e = laser energy density (mJ/cm^2)
p = laser pulse repetition rate (Hz)

The merit of the observer *in situ* state estimator also includes attenuation of process noise and disturbance propagation for improved *in situ* control performance, where estimated subprocess parameter values, $\hat{x}$, are employed in the control algorithm instead of measured parameter values, x. Deviation of x from $\hat{x}$ values constitutes migration of actual relative to estimated *in situ* subprocess states. This detection, in concert with *in situ* references, r, initiates remodeling of the observer feedback controller matrix, E, to achieve convergence of $\hat{x}$ to x. That process of remodeling pole placement constitutes product feature attribution in pursuit of quality maturity. Expert rules governing remodeling for linearized *in situ* subprocess operation are described by Figure 11.8. These rules also accommodate process disorder, such as laser target depletion, by detecting actuator effort that exceeds specified bounds.

11.5 *Spectrometer* in situ *fuzzy control*

Challenges to contemporary materials process control include realizing the potential of *in situ* sensors and actuators beyond apparatus boundaries for process observation and intervention. Increased processing effectiveness is aided by system decomposition into the natural hierarchy of linear and decoupled parameter influences previously presented that link environmental, *in situ*, and *ex situ* subprocesses. Subprocess measurements of mass–momentum–energy state values enable both *ex situ* planner updating and feedback control for processing disorder minimization. Figure 11.9 illustrates the specific measurements employed.

The high kinetic energy involved in laser-ablated material, with wide-ranging target compounding options plus additional chemical sources such as magnetron sputtering, combine to enable versatile stoichiometric material growth capabilities. Achieving product property goals with a uniform deposition rate, in the minority, is assisted by stabilizing boundary conditions through environmental subprocess regulation of process apparatus parameters, feedstock mass flow, and energy input. In the majority, however, *in situ* control of plume species density and deposited film microstructure are the progenitors of processing performance.

Environmental subprocess excitation typically includes excimer laser energy of 300 mJ illuminating a 15-mm^2 target footprint, with a 17-nsec pulse of 248 nm and a 20-Hz rate, in an apparatus vacuum of 10^{-8} Torr. Environmental control references that maximize plume output are synthesized by an *in situ* fuzzy controller employing emission spectrometer sensing of plume chemical spectra to vary laser cavity kilovoltage as the manipulated process variable. This wide bandwidth controller is shown in Figure 11.10 and is capable of pacing process dynamics to maintain peak plume velocity. The slower film growth rate permits deposited microstructure to be acquired by an *in situ* microbalance sensor for control purposes, as defined by equation (11.5).

A principal limitation of proportional-integral-derivative (PID) control is its very low frequency response, especially attributable to the integral term, which relegates these controllers to slow process control applications even though their controlled-variable setpoint following accuracy is high. In contrast, the wideband fuzzy control loop contains no controller low-frequency

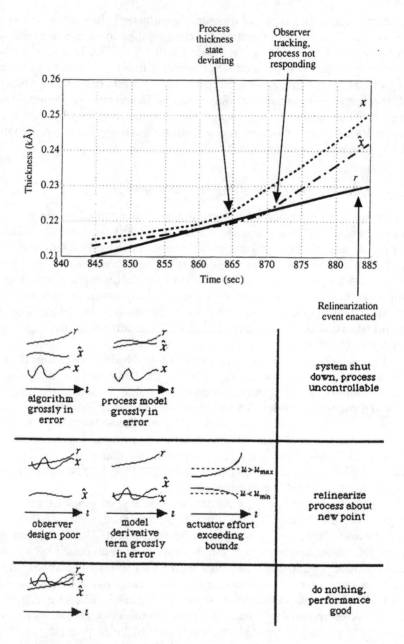

Figure 11.8 Planner remodeling *in situ* observer rules.

response limitation. In fact, minor loop stabilizing error-rate control provides a response widening derivative term enhancing controlled-variable setpoint following for rapidly changing process dynamics.

Fuzzy logic offers *in situ* feedback control of poorly modeled and noisy data process conditions, often encountered in process automation systems,

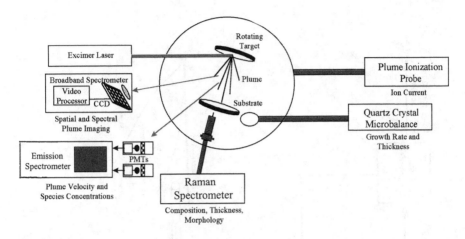

Figure 11.9 Laser deposited material *in situ* sensor suite.

beyond that possible with traditional PID controllers. Process actuation employing triangular input fuzzy membership functions and centroid output defuzzification describes inference based on operator process error and error-rate perceptions, shown in Figure 11.11. This generic fuzzy rule set constitutes a symbolic representation of the plume control algorithm, the overlapping terms of which provide smoothing across the control domain.

11.6 Process data tolerance evaluation

The hierarchical process automation architecture utilized throughout these applications for product manufacturing defines three ascending subprocesses, from environmental to *in situ* to *ex situ* levels, that reveal the principle of decreasing accuracy with increasing complexity for hierarchical ascent. This fundamentally follows Heisenberg's uncertainty principle whereby increasing knowledge is accompanied by decreasing accuracy and vice versa. For this application, *in situ* data accuracies are shown in the feature space of Figure 11.12 to provide indices of system tolerances and product quality capabilities. Note that laser energy accuracy is high at the environmental level, whereas the accuracy for thickness is low at the *ex situ* level. Evaluation of these data follow:

$$t_f = \left[\frac{N_q d_q}{\pi d_f f_c C} \right] \text{ thickness algorithm} \qquad (11.5)$$

where
    t_f = film thickness (cm)
    d_q = quartz density (g/cm^3)
    N_q = crystal frequency constant (Hz/cm)
    d_f = film density (g/cm^3)
    f_c = coated crystal frequency (Hz)
    C = calibration constant (1/cm^2)

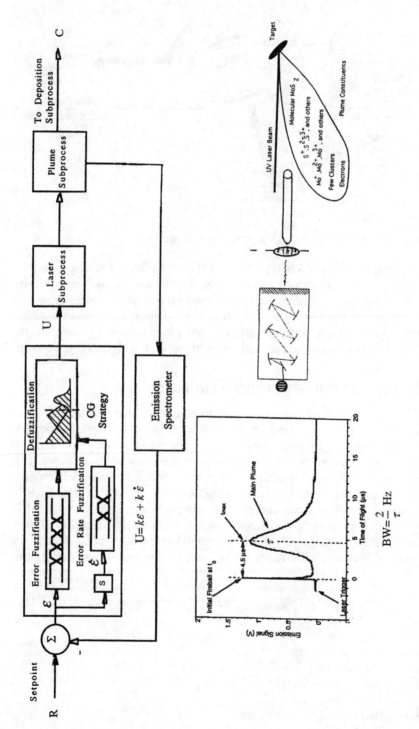

Figure 11.10 Emission spectrometer plume fuzzy control.

		PF	Positive Full		
		PH	Positive Half		
		Z0	Zero		
		NH	Negative Half		
		NF	Negative Full		

		Output				
Error	P	PF	PH	NH	NH	NF
Rate	Z	PF	PH	Z0	NH	NF
	N	PF	PH	PH	NH	NF
		NL	NS	ZE	PS	PL
				Error		

P	Positive
Z	Zero
N	Negative

PL	Positive Large
PS	Positive Small
ZE	Zero
NS	Negative Small
NL	Negative Large

Figure 11.11 Fuzzy logic control rules.

Laser-deposited material process sensing of substrate thickness and thickness rate permits online deposited film microstructure acquisition, using a quartz crystal microbalance positioned in the direct plume plasma without shadowing the substrate. An Inficon XTC sensor indicates thickness, t_f, between 10 and 10,000 Å by crystal frequency changes resulting from deposited mass buildup, according to equation (11.5). Independent offline

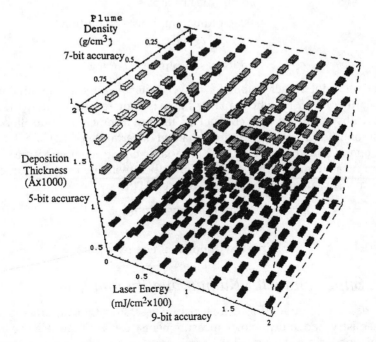

Figure 11.12 Mass–momentum–energy process data.

scanning electron microscope characterization of the sensor film verified a thickness variability of approximately 3%FS, providing a resolution 2^{-n} of 0.03 that approximates 5-bit binary accuracy by the following equation:

$$2^{-n} = 0.03 \text{ thickness resolution} \tag{11.6}$$

$$n = |3.32 \log (2^{-n})|$$

$$= 5 \text{ bits } ex \text{ } situ \text{ thickness accuracy}$$

Optical emission spectroscopy of the plume permits real-time chemical line spectra measurement, employing wideband digitization provided by a 400-megasample f_s oscilloscope, for the example 1.5-μsec width, –3-dB plume amplitude response contained in Figure 11.10. The binary accuracy realized from this sensor data is relevant to both the feedforward *ex situ* planner process observer and plume *in situ* feedback fuzzy controller. Signal representation in terms of step-interpolated binary data, as furnished by an A/D converter to a computer data bus, is evaluated for its 2^{-n} amplitude resolution fraction (referenced to unity full scale) and corresponds to 7-bit binary accuracy. This is described by the following equation:

$$2^{-n} = \frac{\sqrt{2} \; \pi \text{ plume BW}}{\sqrt{5} \text{ sampling} f_s} \tag{11.7}$$

$$= \frac{\sqrt{2} \; \pi \; (2/1.5 \text{ μsec width})}{\sqrt{5} \; 400 \text{ megasamples}}$$

$$= 0.0066 \text{ plume amplitude resolution}$$

$$n = |3.32 \log (2^{-n})|$$

$$= 7.2 \text{ bits } in \text{ } situ \text{ plume accuracy}$$

The ultraviolet laser bolometer power meter possesses a nominal mean nonlinearity of 0.1 %FS whose analog signal is digitized, for input to the laser control subsystem, by one channel of the environmental controller data acquisition interface with a random channel uncertainty of 0.11%FS 1σ. This subsystem is shown in Figure 11.2. The laser energy combined mean and random measurement error of 0.21%FS is then evaluated for its numerical energy measurement resolution 2^{-n} value of 0.0021 by:

$$2^{-n} = 0.0021 \text{ laser energy resolution}$$

$$n = |3.32 \log (2^{-n})| \tag{11.8}$$

$$= 9 \text{ bits environmental energy accuracy}$$

11.7 Superconductor Raman quality analysis

Sensor-based recognition for microstructure assessment, incorporating material chemistry and morphology measurements, can facilitate feature-based determination of product quality. Raman scattering is ideal for the spectroscopy of molecules and crystals, but is a weak effect only 10^{-6} of Raleigh

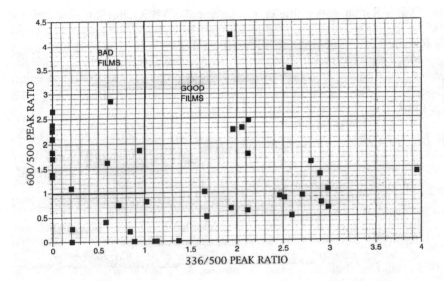

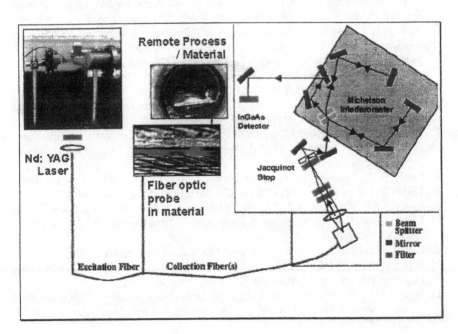

Figure 11.13 Raman spectrometer YBCO quality.

scattering such that near-IR laser excitation and sensitive detection employing photomultipliers are required owing to typical process noise levels. Rotating and vibrating molecules of sensed media result in new frequencies, analogous to inelastic corpuscular collisions, the intensities of which are directly proportional to the concentration of scatterers in the measurement sample.

Offline Raman spectrometer measurement of YBCO film stoichiometry for 42 laser-deposited material production samples is shown in Figure 11.13. Films of Raman wavenumber cm^{-1} shift peak values for ratios 336/600 less than 1 are poor, being oxygen deficient and mostly nonsuperconducting, especially when accompanied by 600/500 values greater than 1. The superconductor 336/500 peak ratio identifying the ideal stoichiometry of $YBa_2Cu_3O_{6.85}$ is 1.5 for films on substrates of $LaAlO_3$.

Bibliography

1. Biggers, R.R., Jones, J.G., Maartense, I., Busbee, J., Dempsey, D., Liptak, D., Lubbers, D., Varanasi, C., and Mast, D., "Emission Spectral Component Monitoring and Fuzzy Logic Control of Pulsed Laser Deposition Processing," *Engineering Applications of AI*, 11(5), 1998, p. 627.
2. Busbee, J., Biggers, R.R., Kozlowski, G., Maartense, I., Jones, J.G., and Dempsey, D., "Investigation of *In Situ* Raman Spectra for Control of PLD of YBCO Thin Film Superconductors," *Engineering Applications of AI*, 13(5), 2000, p. 589.
3. Busbee, J., Igelnik, B., Liptak, D., Biggers, R.R., and Maartense, I., "Towards *In Situ* Monitoring of YBCO T_c and J_e via Neural Network Mapping of Raman Spectral Peaks," *Engineering Applications of AI*, 11(5), 1998, p. 637.
4. Busbee, J., Laube, S.J.B., and Jackson, A.G., "Sensor Principles and Methods for Measuring Physical Properties," *Journal of Materials*, 48(9), 1996, p. 16.
5. Jones, J.G., Biggers, R.R., Busbee, J.D., Dempsey, D.V., and Kozlowski, G., "Image Processing Plume Fluence for Superconducting Thin Film Depositions," *Engineering Applications of AI*, 13(5), 2000, p. 597.
6. King, P.J. and Mandani, E.H., "The Application of Fuzzy Control Systems to Industrial Processes," *Automatica*, 13, 1977, p. 235.
7. Laube, S.J.P., Hierarchical Control of Pulsed Laser Deposition for Manufacture, Ph.D. Dissertation, Electrical and Computer Engineering Department, University of Cincinnati, OH, 1994.
8. Laube, S.J.P., "Pulsed Laser Deposition Improvements by Self Directed Control," USAF Technical Report, WL-TR-95-4079, Wright-Patterson AFB, May 1995.
9. Laube, S.J.P. and Stark, E.F., "Artificial Intelligence in Process Control of Pulsed Laser Deposition," IFAC Symposium AI in Real-Time Control, AIRTC, Valencia, Spain, October 1994.
10. Lubbers, D.P., *In situ* Monitoring and Control of Pulsed Laser Deposition of Superconducting Films, Master's Thesis, Electrical and Computer Engineering Department, University of Cincinnati, OH, 1996.
11. Moore, D.C., Subprocess Control Design Methods for Advanced Materials Processing, Master's Thesis, Electrical and Computer Engineering Department, University of Cincinnati, OH, 1994.
12. Murray, S.P., Digital Control System Design and Demonstration for Pulsed Laser Deposition of Superconducting Films, Master's Thesis, Electrical and Computer Engineering Department, University of Cincinnati, OH, 1994.
13. Park, J. and Woods, D., "Discovery Systems for Manufacturing," USAF Technical Report, WL-TR-94-4008, Wright-Patterson AFB, January 1994.
14. Singh, R.K. and Narayam, J., "Pulse Laser Evaporation Technique for Deposition of Thin Films: Physics and Theoretical Model," *Physical Review*, 41(13), 1990, pp. 8843–8859.
15. Suh, N.P., *Axiomatic Design*, Oxford University Press, Oxford, 2001.

chapter twelve

Neural network directed steel annealing

12.0 Introduction

Steel recrystallization annealing is demonstrated for continuous strip ductility reconstitution. This is achieved via a neural network *ex situ* planner, ANNEAL NET, that has been trained offline from previous strip coupons representing material hardness properties. The network provides direction of a gradient descent algorithm to optimize ten annealing zone temperature values for minimizing strip hardness variance. The annealing process apparatus accommodates a cold-reduced steel strip in motion for thermal modification of its crystalline grain microstructure to within two Rockwell units of variability, which is described by three process automation concurrent engineering linkages. Wheeling-Pittsburgh Steel Corporation made their annealing works available for this implementation, including 3,519 product samples for ANN training.

12.1 Steel annealing concurrent engineering

Thermal annealing for restoration of cold-reduced steel strip ductility is a common steel production process amenable to improvement by means of computationally intelligent processing. Limitations arise from the significant size of the physical apparatus in restricting process modifications, illustrated in Figure 12.1, but opportunities also accrue from the versatility of performance improvement capabilities possible. Improving product value by instituting comprehensive process compensation beyond traditional trial-and-error methods is the objective. In the outcome, three process automation concurrent engineering linkages are realized that benefit processing, as shown in Figure 12.2. Included are improved product property modeling, an ANN-based *ex situ* planner, and *in situ* subprocesses with environmental control execution.

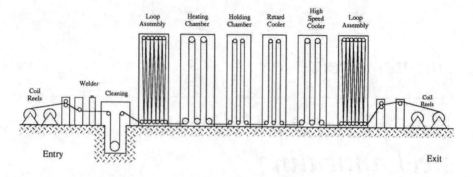

Figure 12.1 Steel annealing process apparatus.

Wheeling-Pittsburgh Steel Corporation made their annealing works available for this implementation, including 3,519 statistically significant previous product annealing data samples for ANN training. Axiomatic parameter decoupling was not considered owing to practical limitations on physical apparatus modifications, and the time-invariant states existing for *in situ* subprocess variables.

12.2 Recrystallization annealing physical properties

Steel recrystallization annealing requires product property modeling capable of defining steel strip ductility from postprocessing sample testing. Cold-reduced steel strip results in grain elongation with increased strength and hardness but a decrease in ductility, which limits its value for essential product forming such as stamping operations. Fortunately, reheating cold

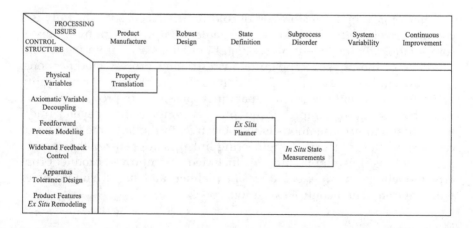

Figure 12.2 Steel annealing concurrent engineering.

rolled steel to a nominal recrystallization temperature of 750°F enables restoration to more ductile steel microstructure properties. However, the principal
challenge of continuous strip ductility reconstitution is minimizing its
end-to-end hardness variability by process control refinement in the application
of specific heat to the strip.

Rockwell hardness evaluation is an *ex situ* mechanical measurement of
strip deformation resistance, applied to test coupons extracted from a typical
strip product of interest, as shown in Figure 12.3. The Rockwell differential
indentation depth measurement is instrumental to evaluation accuracy
because of its insensitivity to coupon dimensions or surface irregularities.

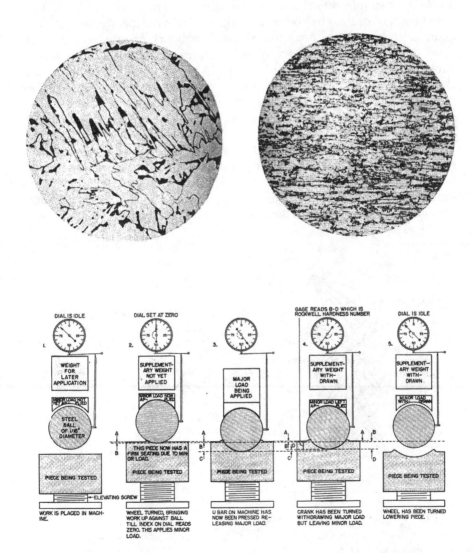

Figure 12.3 Rockwells of originally cast vs. cold reduced steel.

Steel strip annealing has a typical processing goal for Rockwell hardness of 60. The use of excessive test coupons only achieves ANN overtraining, with no additional product improvement.

12.3   *ANNEAL NET* ex situ *planner*

To achieve product goals, production processes require appropriate models to direct controlled variables along ideal process state trajectories. This assumes the existence of capable process and control apparatus as well as necessary material property modeling. Further, two considerations are invoked for simplification of the recrystallization annealing *ex situ* planner: time-invariant process behavior and strip hardness independence from metallurgical composition. An artificial neural network process model designated ANNEAL NET was accordingly structured to provide one output parameter of strip Rockwell hardness for 13 process parameter inputs: strip gauge, strip width, strip speed, and ten zone temperatures. Following ANN training, process zone temperatures are derived by a gradient descent algorithm that provides hardness values of interest based upon accumulated knowledge of the training data.

ANNEAL NET is a feedforward network, with one hidden layer employing backpropagation for training weight values, w, that incorporate a momentum factor, alpha, to accelerate learning. The structure used for this design is shown in Figure 12.4. Backpropagation is a learning procedure for efficiently calculating the derivatives of the output of a nonlinear differentiable system,

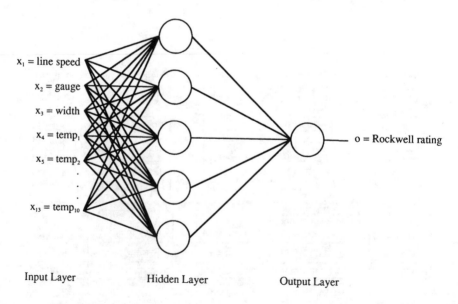

x_1 = line speed

x_2 = gauge

x_3 = width

x_4 = temp$_1$

x_5 = temp$_2$

x_{13} = temp$_{10}$

o = Rockwell rating

Input Layer   Hidden Layer   Output Layer

Figure 12.4 ANNEAL NET model architecture.

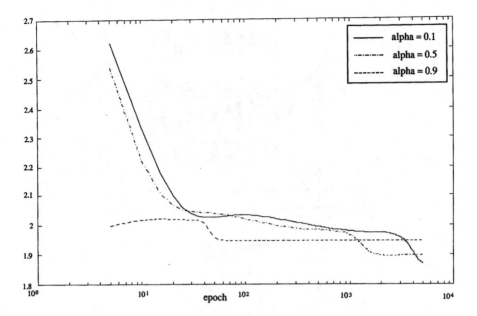

Figure 12.5 ANNEAL NET Rockwell deviation vs. training epochs.

with respect to all inputs and parameters of the system, through calculations that proceed backwards from outputs to inputs. A bipolar sigmoidal activation function, $f(S)$, ensures the network's ability to accommodate the large number of inputs present, where its derivative is implemented as $f'(S) = \lambda/2[1 - f(S)^2]$, with λ regulating steepness. The hidden layer consists of five neurons for this network. Training error is shown employing 3,519 product data samples in Figure 12.5, each consisting of measured hardness, where a deviation error of approximately two Rockwell hardness units is predicted by the network.

Ex situ planner zone temperature determination realizes operating temperature values based on production strip gauge, width, and line speed inputs. These inputs provide convergence to a goal hardness, h, by ANN output hardness, o, using updated estimates employing a gradient descent algorithm. The ten zone temperatures estimated provide environmental setpoints for the annealing process control apparatus. A diagram of the zone temperature evaluation algorithm is shown in Figure 12.6.

A relative cost function error, C, guides this evaluation by encouraging network hardness output changes that decrease C for sample input data and trial zone temperature iterations. Cost function minimization is achieved with a negative direction of the cost gradient through adaptation of ANN weights. Employing a sensitivity derivative enables the gradient of the ANN output to be fed back as updated estimates for backpropagation training.

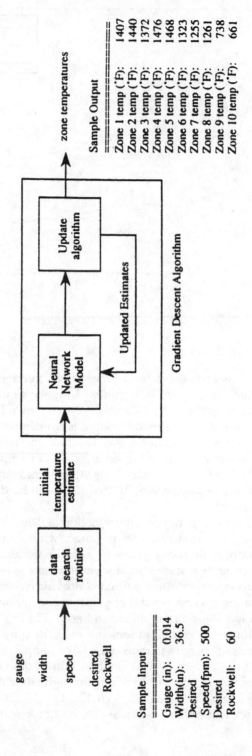

Figure 12.6 ANNEAL NET zone temperature algorithm.

Of interest is a compromise between learning momentum and error minimization, which is realized by the weight iteration equation and plotted for several alpha values in Figure 12.5.

$$f(S) = \frac{1}{1 - e^{-\lambda s}} - 1 \qquad \text{bipolar sigmoid} \qquad (12.1)$$

$$C = \frac{1}{2}(o - h)^2 \qquad \text{cost function} \qquad (12.2)$$

$$\frac{\partial C}{\partial x} = -(h - o)\frac{\partial o}{\partial x} \qquad \text{sensitivity derivative} \qquad (12.3)$$

$$\Delta w(i) = \frac{\partial C}{\partial W(i)} + \alpha \Delta w(i - 1) \quad \text{weight iteration} \qquad (12.4)$$

12.4 Steel annealing in situ control

Figure 12.7 reveals a hierarchical relationship between apparatus and product subprocesses and their control structure. The annealing process apparatus accommodates a cold-reduced steel strip in motion for thermal modification of its deformed crystalline grain microstructure. To recover original strip ductility with minimum Rockwell hardness variability, automatic control of strip speed and ten heating zones are directed by the ANNEAL NET *ex situ*

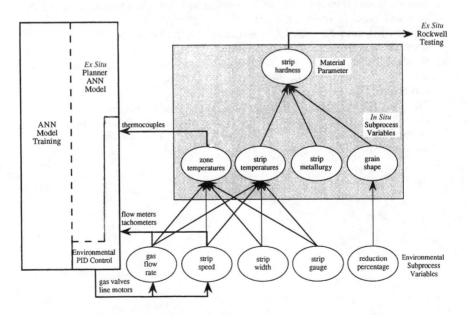

Figure 12.7 Steel annealing process influences.

planner providing temperature setpoints input directly to the environmental controllers. This default to unity *in situ* control arises since temperatures sensed within the ten discrete apparatus heating zones are highly correlated with *in situ* temperatures of the moving strip, because of a favorable ratio of strip surface area to volume for heat transfer, with *in situ* strip temperatures equivalent to their respective zone temperature measurements.

Regulation of *in situ* strip temperatures is consequently achieved by ten environmental feedback control loops for gas flow and one for strip speed adjustment. Although strip width and gauge are crosscoupled *in situ* sub-process influences, their time invariance results in no control iteration difficulty or need for subprocess variable decoupling by means of axiomatic apparatus redesign.

Bibliography

1. Baum, G.A., *Manufacturing Process Controls for the Industries of the Future*, NMAB-487-2, National Academy Press, Washington, D.C., 1998.
2. Gibson, J.R., Sekiguchi, Y., and Rowe, W.H., "Automatic Control System for the Continuous Annealing Line at USS-POSCO," *Iron and Steel Engineer*, 69, 42, 1992.
3. Haykin, S., *Neural Networks*, Macmillan College Publishing, New York, 1994.
4. Hornick, K., "Approximation Capabilities of Multilayer Feedforward Networks," *Neural Networks*, 4, 251, 1991.
5. Jeong, S.H. and Ha, M.Y., "Computer Modeling of the Continuous Annealing Furnace," *Journal of Energy Resources Technology*, 114, 345, 1992.
6. Junius, H.T., "Continuous Annealing of Cold Rolled Steel Sheet — Metallurgical and Economic Aspects," *Iron and Steel Engineer*, 66, 45, 1989.
7. Moody, J. and Darken, C.J., "Fast Learning in Networks of Locally-Tuned Processing Units," *Neural Computation*, 1(2), 281, 1989.
8. Ponton, J.W. and Klemes, J., "Alternatives to Neural Networks for Inferential Measurement," *Computers and Chemical Engineering*, 17(10), 991, 1993.
9. Reddy, R.A. ANNEAL NET: Wheeling-Pittsburgh Steel Hardness Variance Minimization, Master's Thesis, Electrical and Computer Engineering, University of Cincinnati, OH, 1996.
10. Thelning, K., *Steel and Its Heat Treatment*, Butterworths, London, 1984.
11. Zurada, J.M., *Introduction to Artificial Neural Systems*, West Publishing Co., St. Paul, MN, 1992.2

chapter thirteen

X-Ray controlled vapor infiltration ceramic densification

13.0 Introduction

High temperature ceramic matrix composites provide advancement for materials-challenged aerospace components including satellite thrusters, jet engine combustors, and afterburner nozzles. Operating temperature extension to 2,500°F for contemporary engines permits a closer approximation to efficient stoichiometric combustion performance. Chemical vapor infiltration is demonstrated to be a versatile processing method for achieving nonporous microstructure for refractory products. This includes *in situ* x-ray density measurement for reactant vapor infiltration process control, and microwave-heated homogeneous densification employing infrared sensors throughout the component volume. Product property modeling is accordingly implemented by a process production *ex situ* planner to effectively direct chemical vapor infiltration *in situ* subprocess control employing multisensor measurements for enhanced online product characterization.

13.1 Ceramic matrix composite concurrent engineering

Referring to Figure 13.1, three concurrent engineering process control linkages are described for ceramic composite manufacture. The first linkage describes the implementation of product properties by means of apparatus variables. However, this translation is not exact and requires additional capabilities. An essential one of these is the direction of controlled variables along state trajectories by means of a feedforward-modeled *ex situ* planner. The planner defines ideal process production throughout a processing cycle including the attenuation of unmodeled processing dynamics. The third linkage details implementation of multisensor *in situ* subprocess feedback control for mass, momentum,

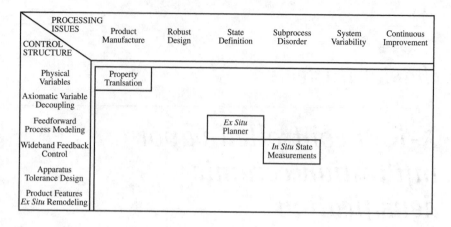

Figure 13.1 Ceramic composite concurrent engineering.

and energy process parameters. Meeting subprocess response with equal control bandwidth is a principal distinction of *in situ* control.

The appreciable parameter crosscoupling between environmental and *in situ* subprocesses prevents elimination of subprocess interactions and controlled variable iteration, resulting in less than robust process performance. That outcome also diminishes the possible benefit of apparatus tolerance design. Realization methods are described for ceramic composite manufacture in the following sections.

13.2 *Combustor product property modeling*

Product manufacture requires identification of apparatus variables in order to define processing sequences capable of realizing product properties of interest. Contemporary ceramic matrix composite fabrication employing chemical vapor infiltration offers nonporous microstructure and preform densification of the product. This is achieved by appropriate feedstock partial pressures and thermal reactions for effective volumetric energy, mass, and momentum transport. Chemical vapor processing is introduced in Chapter 8.2 Product Property Apparatus Variables describing influences that determine product composition and microstructure, including modeling by the Navier-Stokes equations shown. Ceramic matrix composites are useful for both high temperature industrial applications and corrosive chemical environments.

The production of ceramic matrix composite combustors permit aircraft engine internal temperatures to 2,500°F, shown in Figure 13.2, applying methyltrichlorosilane (MTS) chemical vapor infiltration for densification of silicon carbide (SiC) combustor preforms. These products provide high thermal shock resistance with light weight. Essential requirements include determination of real-time profile density for actuation of gas reactant mass flows

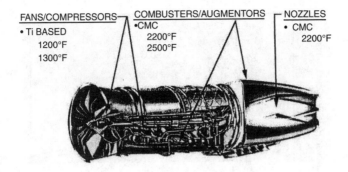

FANS/COMPRESSORS
- Ti BASED
  1200°F
  1300°F

COMBUSTERS/AUGMENTORS
- CMC
  2200°F
  2500°F

NOZZLES
- CMC
  2200°F

Figure 13.2 Combustor product property application.

during manufacture, to aid achievement of uniform densification with acceptable yields absent porosity. In process, gaseous MTS is dissociated at the preform depositing solid SiC with a residue of gaseous hydrochloric acid. This reaction is described by equation (13.1). The goal hydrogen-to-MTS volume ratio during processing is modeled by equation (13.2).

$$CH_3SiCL_3 \rightarrow SiC + 3\ HCL \tag{13.1}$$

$$\frac{\text{Volume H}_2}{\text{Volume MTS}} = \left(1 + \frac{\text{Flow H}_2}{\text{Flow MTS}}\right)\left[\left(\frac{\text{Pressure} \bullet 10^{\frac{1628°k}{\text{Temp}°k}}\text{psi}}{5 \times 10^7 \text{psi}}\right) - 1\right] \tag{13.2}$$

13.3   *Microwave densification* ex situ *planner*

Ex situ-planner-modeled deposition production enables direction of process controlled variables through ideal states to achieve product properties that are described in Section 13.2. This model accordingly describes event-based process execution that defines process behavior throughout a processing cycle by means of *in situ* control reference generation. A primary performance enhancement is attenuation of unmodeled processing dynamic events arising from both internal and external sources of process disorder. That is physically achieved by the combined effectiveness of *ex situ* control of long-time-constant process disturbances and *in situ* control of short-time-constant process variability.

Factorial chemical vapor reactor apparatus modeling by the methods of Chapter 8.2 implies surface densification with progressive interior porosity corresponding to a function of temperature. The greatest uniformity is predicted by the *ex situ* planner at 900°C, (see Figure 13.3) for process conditions that maximize mass transport but can require processing times to 1000 hours. Densification uniformity is aided by volumetric microwave heating instead of preform external radiative heating. With volumetric microwave heating interior temperatures, and hence interior densification, beneficially are not

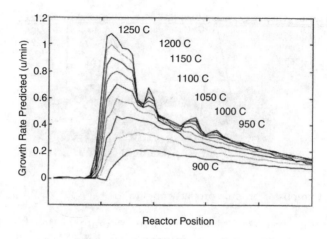

Figure 13.3 Composite deposition rate prediction.

exceeded prematurely by surface densification, as with external heating, owing to the microwaved surface being cooler than the interior since heat can only be lost through the surface. Microwave heating thermal diffusion is described by equation (13.3). Process apparatus are shown by Figure 13.4.

$$k\frac{1}{z}\frac{d}{dz}\,z\,\frac{dT}{dz} = -P \qquad (13.3)$$

k = preform thermal conductivity (W/m °C)
T = temperature °C
P = volumetric microwave power (W/m³)

13.4   X-Ray and infrared in situ control

For *in situ* subprocess control, production planner references are executed over product processing states by *in situ* feedback control loops. This control system is described by the ceramic composite process influences diagram of Figure 13.5 that shows the relationship between subprocesses and their controllers. A further attribution of *in situ* control is multisensor integration which combines heterogeneous process measurements to achieve improved product property characterization generally not available from single sensors.

Meeting *in situ* subprocess response with like closed-loop control bandwidth is also essential for effective processing. In the absence of this equality only environmental variable regulation is possible, with regression to open-loop *in situ* subprocess actuation absent the benefits of feedback control. Fortunately for combustor manufacture x-ray density and infrared temperature measurements, respectively, provide closed-loop *in situ* control of reactant gas mass flow and microwave heating with control bandwidth that equals or exceeds the very slow densification process response.

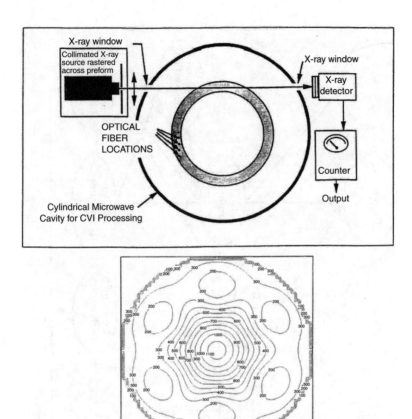

Figure 13.4 Process with microwave power pattern.

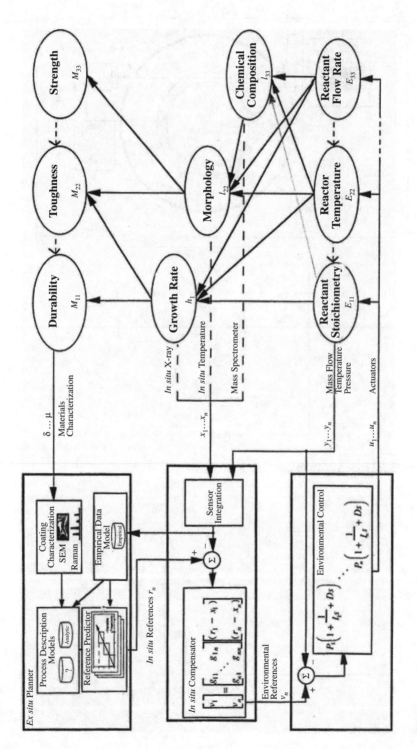

Figure 13.5 Ceramic composite process influences.

Bibliography

1. Allendorf, M.D. and Kee, R.J., "A Model of Silicon Carbide Chemical Vapor Deposition," *Sandia Report* SAND90-8584, June 1990.
2. Brown, P.W. et al., "Transport Phenomena in Chemical Vapor Deposition of Silicon Carbide by the MTS/Hydrogen System," *J. Materials Processing & Manufacturing Science*, Vol. 1, January 1993.
3. Garrett, P.H. et al., "Self-Directed Processing of Materials," *IFAC Engineering Applications of Artificial Intelligence*, Vol. 12, October 1999.
4. Gupta, D. and Evans, J.W., "A Mathematical Model for CVI with Microwave Heating and External Cooling," *J. Material Research*, Vol. 6, 1991.
5. Jensen, K.F., "Chemical Vapor Deposition," American Chemical Society 0065–2393/89/0221, 1989.
6. Jones, J.G., "Intelligent Process Control of Fiber Chemical Vapor Deposition," Electrical Engineering Ph.D. Dissertation, University of Cincinnati, 1997.
7. Morell, J.I. et al., "A Mathematical Model For CVI Volume Heating," *Electrochemical Society*, Vol. 139, 1992.
8. Naslain, R. et al., "The CVI Processing of Ceramic Matrix Composites," *J. de Physique*, Vol. 50, 1989.
9. Palaith, D. and Fehrenbacher, L., "Hierarchical Process Control of Chemical Vapor Deposition/Infiltration," Technology Assessment & Transfer, Final Report, AFRL-ML-WP-TR-1999–4130, July 1999.
10. Palaith, D. "Microwave Joining of SiC," DOE Adv. Materials Concepts Symp., Albuquerque, June 1991.
11. Patankar, S.V., *Numerical Heat Transfer and Fluid Flow*, McGraw-Hill, 1980.
12. Sotirchos, S.V., "Dynamic Modeling Of Chemical Vapor Infiltration," *AIChE Journal*, No. 37, 1991.

Index